Guiding Future Homeland Security Policy

Advances in Homeland Security
Sandra F. Amass, Series Editor
Volume 2

Discovery Park

Guiding Future Homeland Security Policy

Directions for Scientific Inquiry

Edited by

Sandra F. Amass
Purdue University
Alok R. Chaturvedi
Purdue University
Srinivas Peeta
Purdue University

Purdue University Press
West Lafayette, Indiana

Printed in the United States of America

Library of Congress Cataloging-in-Publication Data

Guiding future homeland security policy : directions for scientific inquiry / edited
by Sandra F. Amass, Alok R. Chaturvedi, and Srinivas Peeta.
    p. cm. – (Advances in homeland security)
  Includes bibliographical references and index.
  ISBN-13: 978-1-55753-418-7 (alk. paper)
  1. Terrorism—United States—Prevention. 2. Terrorism—Government policy—
United States. 3. National security—United States. 4. Science and state—United
States. I. Amass, Sandra F., 1967- II. Chaturvedi, Alok R. III. Peeta, Srinivas. IV. Series.
  HV6432.G86 2006
  363.325'15610973–dc22
                              2005038031

Contents

Preface

Welcome to the second volume of the *Advances in Homeland Security* book series. Volume One of the *Advances in Homeland Security* book series presented by the Purdue University Homeland Security Institute provided a critical review of the scientific literature on current topics in Homeland Security. The objective of this volume, *Guiding Future Homeland Security Policy: Directions for Scientific Inquiry* was to provide a venue for emerging topics that do not yet have a strong basis in the scientific literature. Authors were asked to provide commentary on current information in their fields and to provide direction for future scientific research in each area. The Purdue University Homeland Security Institute plans to include additional topics in future volumes of this series as all emerging fields in Homeland security could not be covered in this volume.

A special thanks goes to the authors for providing their works and the editorial group for providing a rigorous review of submitted materials. The assistance of Dr. Eric Dietz and Gina Niemi in preparation of this volume is gratefully acknowledged. Finally, the Institute thanks Thomas Bacher and Dr. Margaret Hunt of the Purdue University Press for their hard work and support, enabling this book series to be possible.

Guiding Future Homeland Security Policy

Chapter 1

Introduction

RUTH A. DAVID
Analytic Services Inc.

> The world changed on September 11, 2001. We learned that a threat that gathers on the other side of the earth can strike our own cities and kill our own citizens. It's an important lesson; one we can never forget. Oceans no longer protect America from the dangers of this world. We're protected by daily vigilance at home. And we will be protected by resolute and decisive action against threats abroad. (Bush 2002)

On September 11, 2001, American citizens as well as friends and allies around the world were stunned by the sheer audacity of the terrorists who attacked our nation. Using commercial aircraft—loaded with passengers—as guided missiles to kill thousands of people and destroy key symbols of our society was an unfathomable act of violence. Observers were rightfully horrified; some were astounded, whereas others saw the attacks as an inevitable demonstration of our nation's vulnerability to terrorism.

The events of 9/11 are often referred to as America's wake-up call. But the so called global war on terror did not begin on 9/11, nor will it end with the demise of the perpetrators of those attacks. Terrorism is a tactic as old as humanity; in recent times we have observed an expansion in terrorism as an asymmetric strategy increasingly employed by disaffected individuals and groups.

1

Concerns about asymmetric threats moved to center stage for our national security community with the end of the Cold War standoff between the United States and the Soviet Union. Throughout the 1990s, numerous commissions and panels called attention to growing threats from transnational terrorist organizations—at home as well as abroad. Their warnings were punctuated by events such as the 1993 bombing of the World Trade Center, the near-simultaneous attacks on two U.S. embassies in Africa in 1998, and the crippling of the USS *Cole* in 2000.

In 1998, the United States Commission on National Security/21st Century was tasked to help create a national security strategy better matched to the emerging threat environment. In the commission's Phase I report, published in September 1999, the number one conclusion was that "America will become increasingly vulnerable to hostile attack on our homeland, and our military superiority will not entirely protect us" (U.S. Commission on National Security/21st Century 1999). Their final report, published in January 2001, was remarkably prescient in stating "a direct attack against American citizens on American soil is likely over the next quarter century" (U.S. Commission on National Security/21st Century 2001).

But prior to 9/11 little progress was made in addressing the growing concerns about our nation's inability to thwart terrorism. Only in the aftermath of the 9/11 attacks was there broad-based acknowledgement that key instruments of our national power were inadequate for this changed world.

Diplomacy is of limited value when there is no reciprocal governing body with whom to negotiate. Our intelligence apparatus failed to warn of the impending attack—highlighting the fault lines between domestic and foreign authorities as well as the vulnerabilities inherent in our open society. Military operations subsequent to 9/11 reduced the threat from Al-Qaida, but also demonstrated that the military capabilities that conferred 20th-century superpower status are not well matched to the first war of the 21st century—the global war on terror. And the damage inflicted on our nation's economy by the 9/11 attacks extended well beyond the lives lost and property destroyed on that day.

But perhaps the most stunning aspect of the 9/11 attacks was the fact that the weapons employed were owned and operated by our own private industry and the targets attacked were predominantly populated by civilians. The events on that day broke all the traditional rules of warfare and shattered the perception that 20th-century security solutions are adequate for 21st-century threats. The wake-up call provided by 9/11 became a call to action for our nation—as well as for friends and allies around the world.

Much has been accomplished since 9/11. In the immediate aftermath the president established the Office of Homeland Security and chartered it "to develop and coordinate the implementation of a comprehensive national strategy to secure the United States from terrorist threats or attacks" (Executive Order 2001). Military operations were launched in an effort to prevent additional threats from reaching our borders. Congress moved rapidly to enhance law enforcement investigatory tools, passing the Uniting and Strengthening America by Providing Appropriate Tools Required to Intercept and Obstruct Terrorism Act—also known as the USA PATRIOT Act. Similarly, the Aviation Security Act was enacted to address perceived weaknesses in aviation security.

The first-ever National Strategy for Homeland Security was published by the Office of Homeland Security in July 2002 to provide a roadmap for the path ahead. Legislation signed into law later that same year, the Homeland Security Act of 2002 initiated a massive reorganization of the federal government and established the Department of Homeland Security.

Work by the National Commission on Terrorist Attacks Upon the United States (also known as the 9/11 Commission), which was established in 2002, spawned legislation that addresses recognized deficiencies in our nation's intelligence apparatus. In December 2004 the president signed into law the Intelligence Reform and Terrorism Prevention Act, which again had significant organizational implications for our federal government.

In the years since 9/11, significant new legislation has been enacted, new protective measures have been put into place, new organizational structures have been established, and new partnerships have been forged. But formidable challenges lie ahead as we deal with the dangers of this changed world. Sustained success will require a balanced approach—solutions must be robust against an uncertain and evolving threat, viable in a competitive global economy, and acceptable in the eyes of our citizenry.

The United States recognized early that efforts to secure our nation against terrorism require the active engagement of both the public and private sectors—homeland security is not the sole purview of the federal government. Many of the resources vital to our ability to respond to terrorist attacks are under the command of state, local, or tribal governments. The vast majority of our nation's critical infrastructures are owned and operated by the private sector. Of equal importance are international partnerships; homeland security solutions must engage the global community in the global war on terror.

During the Cold War, our nation's deterrence strategy was based largely on maintaining the balance of power; that strategy was successful—we emerged victorious. But in this changed world, asymmetric options tilt the balance of power—21st-century adversaries may have more in common with Al-Qaida than with the former Soviet Union.

To Prevail: An American Strategy for the Campaign Against Terrorism, published in the aftermath of 9/11, suggested that the strategic mantra for the future should be the *power of balance* rather than the *balance of power* (Campbell and Flournoy 2001). That is, deterrence in the 21st century will require an evolving suite of capabilities that hedges our bets against thinking adversaries who are equipped by an infinite array of asymmetric options. But investment in such capabilities must be balanced against the impact—societal as well as economic—that could accrue from their implementation.

The National Strategy for Homeland Security identified the need to foster a flexible response to terrorism—recognizing that "our terrorist enemies can strategically adapt their offensive tactics to exploit what they perceive to be weaknesses in our defenses" (Office of Homeland Security 2002). Homeland security solutions must anticipate threats not yet observed—we must think anew.

The national strategy also identified the need to "make difficult choices about how to allocate resources against those risks that pose the greatest danger to our homeland" (Office of Homeland Security 2002). Resources available for securing our homeland are finite even though perceived threats are not. Investment priorities must be guided by the principles of risk management—by thoughtful assessment of threats, coupled with an understanding of likely targets and their vulnerabilities to those threats, and integrated with an evaluation of the consequences should a given attack occur.

As acknowledged in the national strategy, "protecting the homeland from terrorist attack is a permanent mission" (Office of Homeland Security 2002). Homeland security solutions must be sustainable over the long run. Only then will we regain the asymmetric advantage currently held by those who employ terrorism against us.

▶▶ Thinking Anew

> The terrorists who seek to attack us are not ready to concede defeat. Rather, they appear determined to adapt their methods to create new threats to our homeland. In order to meet this evolving threat, we must be willing and eager to think anew as well. (Chertoff, March 2005)

Just as on September 10, 2001, we would not have expected a hijacker to turn a commercial airplane into a guided missile, on September 12, 2001, we did not envision exploding shoes as a threat to aviation. Nor did we anticipate the use of our United States Postal Service as a delivery system for anthrax-laden letters or the use of cell phones to remotely detonate improvised explosive devices. While we must be vigilant against these and other observed threats, we must balance that vigilance with equal passion for thinking anew.

We must acknowledge the capacity of our adversaries to observe our defensive actions and to adapt their methods accordingly—exploiting the myriad asymmetric options afforded by terrorist tactics. Terrorist networks are not geographically anchored; membership and support infrastructures often span national borders—including our own. Terrorists need not rely on the conventional tools of war; their objectives can be accomplished by employing materials and information that often is readily available in the global marketplace. And because terrorists are not bound by the traditional rules of warfare, their selection of weapons, delivery systems, and targets is virtually unlimited.

One of the most problematic aspects of the global war on terror is that many, perhaps most, of the apparently attractive targets are owned and/or operated by the private sector. Most businesses have by now accepted the need to defend against ongoing threats posed by hackers and computer viruses. Many recognize that they must prepare for accidents and other disasters because they on occasion find themselves in the path of a hurricane or victim of a power blackout. Virtually all businesses have measures in place to protect against crime and fraud.

But prior to 9/11 few corporate executives thought of their business as a potential terrorist target. The private sector, too, must think anew. Policies and processes designed to ensure business continuity and resilience in the face of natural disasters or accidents can also mitigate the consequences of a terrorist attack. Similarly, access control and information management systems designed to protect proprietary business information for competitive advantage can also protect against exploitation by terrorist organizations. But the policies, processes, and systems must be designed and implemented with these issues in mind—corporate executives must now factor the threat of terrorism into their planning processes.

Of equal concern is the fact that American citizens—as well as visitors to our homeland—are on the frontlines in the global war on terror. This is not a war waged entirely on foreign soil, nor is it a war in which an enemy is readily distinguishable from an innocent bystander. Prevention of future attacks depends significantly upon our ability to identify and interdict terrorists before they have an opportunity to act.

The requisite capabilities will inevitably impact the lives of nonterrorists as well. The events of 9/11 stimulated immediate changes in passenger security screening in our nation's airports; the subsequent shoe bomber stimulated yet another round of inspection. Access control procedures have since been modified in many other facilities throughout the public and private sectors. Additional measures are likely on the horizon as we grapple with growing security concerns in areas such as transit security.

These and related changes impact many individuals on a daily basis—our citizens, too, must think anew. They must engage in productive debate regarding contentious policy issues and embrace novel solutions that strike the right balance—protecting lives while also protecting our way of life. Solutions for securing our homeland against terrorism can—and must—protect the individual freedoms we hold dear. But at the same time we must acknowledge that the world changed, and that 21st-century threats demand 21st-century solutions.

The National Strategy for Homeland Security recognized that "just as science and technology have helped us defeat past enemies overseas, so too will they help us defeat the efforts of terrorists to attack our homeland and disrupt our way of life" (Office of Homeland Security 2002). Currently available technologies will yield valuable solutions if effectively integrated. Significant technological innovation will be needed in many areas. Some of the challenges we face will likely require the invention of technologies not yet envisioned. In every case, however, the value of a technology-based solution is determined by the effectiveness of its implementation—by its integration into a complex systems environment that includes guiding policies, operational processes, and well-trained personnel.

As we think anew—as we design new ways to thwart terrorist threats—we must be mindful of the policy, economic, and societal implications of their implementation. We must engage not only those who will effect the solution but those who will be affected by it. We must be innovative in our approaches, thoughtful in our investment choices, and pragmatic in our implementation of new solutions.

Managing Risk

Risk management is fundamental to managing the threat, while retaining our quality of life and living in freedom. Risk management can guide our decision-making as we examine how we can best organize to prevent, protect against, respond and recover from an attack. (Chertoff, April 2005)

The National Strategy for Homeland Security established three objectives to help set priorities for our nation's decision makers (Office of Homeland Security 2002):

1. Prevent terrorist attacks within the United States
2. Reduce America's vulnerability to terrorism
3. Minimize the damage and recover from attacks that do occur

These objectives implicitly establish a risk management framework to guide decision making.

Risk is traditionally viewed as a function of three parameters: threat, vulnerability, and consequence. By focusing first on prevention of terrorist attacks, we seek to reduce the threat component of the risk equation. The second objective maps directly to the vulnerability component of risk. The third objective, which acknowledges that in spite of our best efforts we will eliminate neither all threats nor all vulnerabilities, focuses on reducing the consequences in the event that an attack does occur.

Application of risk management principles to guide decision making in the face of uncertainty is inherently logical. A well-established methodological basis for risk management is already in use by many organizations throughout the public and private sectors. The technique is designed to focus resources on thwarting events that would have the most severe impacts should they occur—either to prevent the event from occurring, to reduce the system's vulnerability to such an event, or to prepare in advance to respond such that the consequences are rapidly mitigated and full recovery is assured. The risk management philosophy accepts that such an approach will never yield perfect results—residual risk always exists.

The use of risk management principles to guide decision making as we allocate finite resources to secure our nation against uncertain and evolving terrorist threats is also inherently logical. In fact, it is the only viable approach given the enduring character of our national objectives. But implementation of risk-based decision making is fraught with challenges—some of which are unique to the complexities of the homeland security mission.

Implicit in the risk management framework is an assessment of the likelihood that a particular event—or attack, in the case of terrorism—will occur. Given the diversity in potential types of attacks as well as prospective targets, it will be difficult to gain consensus regarding which events are the more likely.

The absence of warning prior to 9/11 is often attributed to a failure of imagination on the part of our nation's intelligence apparatus. It is evident that we must think anew and envision novel tactics that may be employed by terrorists. But we must moderate that imagination with a disciplined assessment of the relative risks to our nation that stem from those tactics.

It is a relatively simple matter to conjure up doomsday scenarios of attacks that are entirely within the realm of possibility and even to estimate at least the direct consequences of such an attack. It is far more difficult, however, to make a priori judgments regarding the likelihood of occurrence. But such judgments are integral to implementation of the risk management methodology.

Our nation provides a target-rich environment in the sense that we cannot protect everyone and everything against all imaginable terrorist attacks—any attempt to do so would bankrupt our economy and destroy our society. As the owners of potential targets make choices about which vulnerabilities to reduce or eliminate, they must carefully weigh both the likelihood that the vulnerability will be exploited and the consequences of its exploitation.

Lessons from 9/11 should not suggest the construction of all new buildings such that they can withstand the impact of a commercial airliner. A more meaningful lesson learned is that hijackers of commercial aircraft may intend suicide rather than escape. Thus, post-9/11 investments designed to keep future hijackers out of cockpits reduce an observed vulnerability and increase the odds of preventing a repeat of the 9/11 scenarios.

An additional measure of complexity in assessing homeland security–related risks is introduced by the multiple stakeholder perspectives regarding consequences. There are many dimensions to consequence and many stakeholder populations that are impacted—either directly or indirectly—by an attack. Some consequences are quantifiable—for example, lives lost or buildings destroyed—some are not. Even though we can count the lives lost due to an attack, it is far more difficult to assess the related economic and psychological impacts.

The consequences of the 9/11 attacks extended well beyond the individuals and facilities inside the ground zero perimeters on that date. In the aftermath, lingering behavioral changes had a widespread impact on the aviation industry as well as the tourism industry because many individuals chose vacation options that did not require them to board a commercial airliner. While such consequences are difficult to quantify a priori, a lesson learned from 9/11 does suggest that more effective communication of risk—hardening of the psychological target—may help mitigate the consequences in the event of future terrorist attacks.

Many experts view a catastrophic biological attack as inevitable—a matter of when, not if. Within the spectrum of potential bioterrorism threats, however, the sheer diversity of pathogens makes prioritization of preventive measures difficult—particularly given that many deadly pathogens exist in nature. Equally challenging is the development of vaccines that would reduce our vulnerability to bioterrorism. The private sector is the logical developer of most vaccines, but the business case that would stimulate such investment is not yet clear, in spite of legislative efforts. Given the likely delayed onset of symptoms, even detection of a bioattack is problematic; the first indications may be spotted in physicians' offices days after the exposure occurred and in geographically dispersed locations.

Given the complexity of the bioterrorism challenge, it seems evident that an integrated strategy for managing the associated risks is desperately needed. Development and implementation of such a strategy will require collaboration and cooperation among disparate stakeholder communities spanning the public and private sectors—each with a stake in the risk management equation. The same is true in many other areas.

Effective implementation of risk management strategies requires systems thinking—and planning—in multiple dimensions. Investment strategies must address the diverse spectrum of already observed threats while anticipating the emergence of new threats. Hardening cockpit doors against hijackers did not deter the subsequent "shoe bomber," nor will vaccines designed for anthrax protect against smallpox.

Risk management strategies must address the interdependencies inherent in our networked economy. The telecommunications and cyber infrastructures underpin virtually every aspect of our society, but those infrastructures are reliant upon the power grid, and the power grid depends on timely delivery of fuels via the transportation sector. Individual businesses rely on these and other infrastructures, as well as suppliers, customers, and employees. Risk management strategies must account for these interdependencies.

Robust solutions for the homeland security mission will integrate diverse stakeholder perspectives—including those from stakeholder communities not accustomed to thinking of themselves as part of a security apparatus. Robust investment strategies will be supported by a policy regime that enables implementation of solutions that emerge. Risk management principles provide the framework within which the dialog among stakeholders must occur and investment priorities must be established.

Building Sustainable Solutions

We must lay out a vision of homeland security that is sustainable over the long run—a vision that balances durable and comprehensive security with the American way of life ... (Chertoff, April 2005)

The strategic challenge ahead is to secure our nation against both observed and emergent threats—without bankrupting our economy or destroying our society. Sustainable solutions will be embedded into the fabric of our daily lives and will deliver benefits beyond securing our homeland against terrorist threats. The most valuable solutions will deliver dual benefits—enhancing our homeland security posture while simultaneously enhancing daily operations.

The dual benefit concept is not new, but neither is it simply a retread of the dual use concept coined by the defense establishment to encourage development of technologies with commercial, as well as military, utility. Dual use focused on reducing unit cost by increasing market size—stimulating the development of technologies of value to multiple users. Dual benefit increases the value to *every* user by delivering simultaneous benefit to missions executed in tandem.

Many potential dual benefit solutions exist—and more are emerging every day. Border management systems that improve knowledge about incoming cargo shipments will speed commerce and also make it more likely that we will identify, in advance, suspect cargo. Critical infrastructures that are resilient to accidents and natural disasters are also more resilient in light of terrorist attacks. Equipping first responders to effectively respond to accidents involving hazardous chemicals also improves their ability to respond to a chemical attack. Miniature sensors embedded in our food supply to detect accidental contamination or spoilage could also warn of intentionally introduced toxins. Authentication techniques designed to protect individuals against identity theft could also help identify terrorists in our midst.

Dual benefit solutions provide return-on-investment beyond a terrorism-focused risk management framework. For the private sector, dual benefit solutions bolster the business case that underpins their investment decisions. Similarly, for the American public, dual benefit solutions provide discernable benefits on a daily basis. Solutions designed to simultaneously protect life as well as our way of life are likely to be more acceptable to our citizens than solutions designed solely to combat the threat of terrorism—particularly as memories of 9/11 fade from our collective consciousness.

The National Strategy for Homeland Security identified as one of eight guiding principles the need to "seek opportunity out of adversity" (Office of Homeland Security 2002). The ultimate irony would be for the threat of terrorism to spawn solutions that make us better as a nation—solutions that simultaneously address other societal problems such as crime and health care while also affording protections against terrorist threats.

Homeland security solutions designed to deliver daily benefits while still addressing the enduring challenges of the homeland security mission enable us to exploit the opportunities inherent in the adversity we face. Such solutions—dual benefit solutions—are far more likely to be sustainable over the long run.

Regaining the Asymmetric Advantage

An array of homeland security solutions that is equal to the sum of its parts is not affordable—our vulnerabilities are too great and the asymmetric options available to terrorists are too diverse. Instead, we must create a national architecture for the homeland security mission—an integrating framework that enables us to more effectively prevent and protect against terrorist attacks and, failing that, enables us to rapidly respond and fully recover.

Our approach must be grounded in the principles of risk management yet sufficiently flexible to cope with a diverse and ever-changing threat environment. We must think anew to effectively counter innovative adversaries as well as to harness our nation's own capacity for innovation. We must effectively mitigate today's risks and yet devise solutions that are sustainable for years to come.

A national homeland security architecture—integrating the complex system-of-systems required to implement homeland security solutions—will enable us to hedge our bets against uncertainty. It will provide the agility needed to counter thinking adversaries and emerging threats. And it will shift the asymmetric advantage to those individuals and nations—worldwide—who fight against terrorism in the global war on terror.

The National Strategy for Homeland Security provided our first roadmap to securing our nation against evolving terrorist threats. But it also highlighted the need to engage the entire nation in the journey ahead. The chapters of this book are intended to contribute to the intellectual foundation for this nascent mission and to help stimulate the necessary national—and international—dialog and debate as we embark upon this journey to regain the asymmetric advantage.

References

Bush, G. W. 2002. As quoted in Strategy for Homeland Defense and Civil Support. Washington, D.C.: Department of Defense, June 2005.

Campbell, K. M., and M. A. Flournoy, eds. *To Prevail: An American Strategy for the Campaign Against Terrorism*. Washington, D.C.: CSIS Press.

Chertoff, M. March 2, 2005. Statement by Secretary of Homeland Security Michael Chertoff. House Appropriations Homeland Security Subcommittee.

Chertoff, M. April 26, 2005. Remarks by Secretary of Homeland Security Michael Chertoff. Center for Catastrophic Preparedness and Response and the International Center for Enterprise Preparedness.

Executive Order 13288. 2001. Establishing the Office of Homeland Security and Homeland Security Council. Washington, D.C.: White House.

Office of Homeland Security. 2002. National Strategy for Homeland Security. Washington, D.C.: White House.

U.S. Commission on National Security/21st Century. 1999. "New World Coming: American Security in the 21st Century." US Commission on National Security/21st Century, http://govinfo.library.unt.edu/nssg/Reports/reports.htm (accessed August 9, 2005).

U.S. Commission on National Security/21st Century. 2001. "Road Map for National Security: Imperative for Change." U.S. Commission on National Security/21st Century, http://www.nssg.gov (accessed August 9, 2005).

Chapter 2

Interagency Relations Regarding Agroterrorism and Agrosecurity

STEVE CAIN AND ABIGAIL BORRON
Purdue University

Proactive interagency coordination will aid in response efforts, whereas a lack of coordination will hinder the progress of necessary actions in the event of a major agricultural disaster. Specifically, steps taken within agencies to provide standard operating procedures (SOPs) that enhance an agency's response, as well as interagency response, are critical to successful outcomes.

Papers and research reports have mentioned interagency relations and agroterrorism, along with associated areas of weakness. However, none have really focused on a detailed strategy that should be followed. Instead, interagency relations in respect to agroterrorism are usually left as a problem with no clear-cut solution.

No state had reported significant gains in interagency relations during the previous year in a 2002–2003 summary of state reports (National Animal Health Emergency Management System 2003). Information that was shared included:

- Alabama State Veterinarian Tony Frazier, DVM: "It is difficult for associated agencies to understand the vulnerability of agricultural bioterrorism. This creates confusion when trying to organize cross-agency training exercises" (National Animal Health Emergency Management System 2003).

- Florida Chief of the Bureau of Animal Disease Control Ashby Green, DVM: "One of our biggest challenges is achieving regular communication, cooperation, and interagency coordination at the local, state, and national levels" (National Animal Health Emergency Management System 2003).

- Montana Department of Livestock, Dr. Arnold Gertonson: "The greatest deficit is at the local level where there has been little or no coordination with local officials or training of local veterinarians" (National Animal Health Emergency Management System 2003).

Many experts agree that interagency relations are critical to the success of a natural or intentional introduction of an agent that would adversely affect agriculture. However, to date, little research material has been published specifically about interagency relations in an agricultural homeland security event. Henry S. Parker observes that "little attention has been given to agricultural biowarfare and bioterrorism or to the roles and responsibilities of the public and private sectors in deterring and responding to potential attacks" (Parker 2002).

The reports in the following list do not focus on interagency issues; however, the reports do provide some insight on federal interagency issues. If reports mention local and state agencies, often it is in reference to how little preparation they have done specifically for agriculture.

- "Agricultural Bioterrorism: A Federal Strategy to Meet the Threat" (Parker 2002)
- "Agroterrorism: Threats and Preparedness" (Monke 2004)
- "Agroterrorism: What Is the Threat and What Can Be Done about It?" (RAND National Defense Research Institute 2003)
- "Bioterrorism: A Threat to Agriculture and the Food Supply" (Dyckman 2003)
- "Countering Agricultural Bioterrorism" (Committee on Biological Threats to Agricultural Plants and Animals, National Research Council 2002)

Researchers concluded, "As of the spring of 2002, no publicly available, in-depth, interagency or interdepartmental national plan had been formulated for defense against intentional introduction of biological agents directed at agriculture" (Committee on Biological Threats to Agricultural Plants and Animals, National Research Council 2002).

The National Incident Management System (NIMS) integrates effective practices in emergency preparedness and response into a comprehensive national framework for incident management. To an extent, NIMS will assist local, state, and federal authorities in an agricultural bioterrorism response. However, assistance will be granted only if certain actions, to be explained later in this chapter, are taken. Addressing interagency issues is complicated, time consuming, and often risky in terms of delving into another agency's territory.

"Understanding government roles and responsibilities in preparing for and responding to acts of terrorism is complicated by the fact that a plethora of federal, state, and local agencies and programs have important, often overlapping, responsibilities for activities that are directly applicable to terrorism" (Parker 2002). Moreover, individual agencies are rarely familiar with roles and protocols of other agencies. Parker advocates that the United States Department of Agriculture (USDA) formally be recognized as the leader for developing a comprehensive agricultural counterterrorism plan. "There are likely to be objections to a USDA-led, stand-alone program to combat agricultural bioterrorism. Other federal agencies may have territory concerns, especially where overlapping jurisdictions already occur (for example, food safety responsibilities in the Food and Drug Administration at the Department of Health and Human Services ...)" (Parker 2002). Although more commentary exists regarding interagency relations in areas outside of agriculture, progress in these areas often seems to be no farther along than that of agricultural agencies.

A white paper was produced that focused on "Ensuring successful collaboration between government agencies and the health care industry to prevent and respond to acts of terrorism" (Kaiser Permanente Health Care Continuity Management Team 2002). The team concluded that, "To effectively respond to a terrorist event, public and private entities need ongoing multidisciplinary public health/terrorism/disaster preparedness planning. Many communities lack a pre-established community health care model for responding to an infectious disease or terrorist incident that exceeds a single hospital's capacity for response" (Kaiser Permanente Health Care Continuity Management Team 2002). The team added, "To prepare for terrorist attacks, ongoing and standardized training is needed for first responders, health care workers in general, and the public at large" (Kaiser Permanente Health Care Continuity Management Team 2002).

All disasters are local. But the level of a response may be different between an agricultural and nonagricultural disaster. Generally, nonagricultural emergency responses are handled using local resources. Assistance at the state and federal levels tends to occur only after the local resources have been depleted.

Similarly, some agricultural disasters may necessitate only a local response. If an agricultural disaster is quickly determined to be nonthreatening—not likely to spread, cause a major food health threat, or cause terror—then the response primarily will be local. For example, cattle at a dairy operation accidentally may have been contaminated with a pesticide. In this case, a local veterinarian and possibly a sheriff would respond to this disaster.

In contrast, the response to a major threatening agricultural disaster immediately will go beyond the local level. More than likely, the response will be state or federal. In most cases, the first detector of the problem, possibly a veterinarian or an extension agent, will call upon state and federal agencies for the initial biosecurity response rather than utilizing local resources. The nonlocal response is required because the biosecurity measures and possible diagnostic processes are regulated at the state or national level. The response to agricultural disasters will vary but likely will include plans to control or eradicate the problem. Local-level responders may still be involved. Additional emergency responders, such as law enforcement, will become involved if the response plan requires isolating large areas, such as a three-mile-radius around the infected plants or animals. Many states are exercising their plans to control or eradicate a bioterrorist-induced event, but few agencies that become involved truly understand the nature of an agricultural event.

RAND Corporation, a nonprofit research organization addressing challenges that face public and private sectors, has been at the forefront of studying agricultural bioterrorism. RAND suggested that "Over the longer term, additional effort should be directed toward standardizing and streamlining food-supply and agricultural safety measures within the framework of a single, integrated strategy that cuts across the missions and capabilities of federal, state, and local agencies. An effort such as this would help to unify the patchwork of largely uncoordinated bio-emergency preparedness and response initiatives that now exist. Integrating agriculture and food safety measures would also reduce jurisdictional conflicts and eliminate unnecessary duplication of effort" (RAND 2003).

Maybe help is on the way. One analyst in agricultural policy wrote, "In terms of protecting critical infrastructure, agriculture was added to the list in December 2003 by Homeland Security Presidential Directive 7 (HSPD-7), 'Critical Infrastructure Identification, Prioritization, and Protection.' . . . These directives instruct agencies to develop plans to prepare for and counter the terrorist threat. HSPD-7 mentions the following industries: agriculture and food; banking and finance; transportation (air, sea, and land, including mass transit, rail, and pipelines); energy (electricity, oil, and gas); telecommunications; public health; emergency services; drinking water; and water treatment" (Monke 2004).

More significant recognition came on January 30, 2004, when the White House released Homeland Security Presidential Directive 9 (HSPD-9),

"Defense of United States Agriculture and Food" (Bush 2004). This directive established a national policy to protect against terrorist attacks on agriculture and food systems. HSPD-9 generally instructs the secretaries of Homeland Security (DHS), Agriculture (USDA), and Health and Human Services (HHS), the administrator of the Environmental Protection Agency (EPA), the attorney general, and the director of central intelligence to coordinate their efforts to prepare for, protect against, respond to, and recover from an agroterrorist attack. In some cases, one department is assigned primary responsibility, particularly when the intelligence community is involved. In other cases, only the USDA, HHS, and/or EPA are involved regarding industry or scientific expertise.

Fifteen months later, the United States Government Accountability Office gave an overall favorable report on agricultural preparedness to Congressional requesters. However, a few interagency relations issues were noted. The report stated, "there are also management problems that inhibit the effectiveness of agencies' efforts to protect against agroterrorism. For instance, since the transfer of agricultural inspectors from USDA to DHS in 2003, there have been fewer inspections of agricultural products at the nation's ports of entry." The report warns of little sharing of the results of state exercises. These exercises may be the best method to test interagency relations and learn how to address them at the state and local levels. Sharing that information could be enhanced with the Department of Homeland Security Information Network (United States Government Accountability Office 2005).

Technology plays an important role in interagency relations. The Internet will create better connectivity between agencies during all phases of disaster: mitigation, preparedness, response, and recovery. Also, technology will allow greater abilities to capture data due to emerging portable electronic capabilities and remote access to databases. Advances in technology will also improve first responders' ability to process and analyze data. For example, in 2005, plant pathologists, farm producers, and the public at large had the ability to watch the movement of soybean rust as it impacted the U.S. soybean crop. A weather-report-like map showing real-time infestation of soybean rust was developed using advances in data collection and database management. Use of technology helped producers and advisors make better crop biosecurity decisions. Use of technology could also help agencies better understand agrobiosecurity issues and, therefore, improve interagency issues.

Technological advances can also create disadvantages. Improvements in technology could potentially create gaps between agencies because technological tools and training are funded differently among agencies. Information overload can result because individuals tend to be subjected to large amounts of data. Finally, new technology can create frustration and time-resource issues as first responders—who may not be properly trained with the technology—grapple with data collection during the pressures of the response phase of a disaster.

Progress is being made to improve the levels of biosecurity within facilities and to improve our ability to detect and diagnose potential problems. However, little has been done to move interagency relations forward in regard to agroterrorism.

▶▶ Roundtable Exercise to Investigate Interagency Relations

The authors have reported on interagency relations and agroterrorism as a result of USDA Animal and Plant Health Inspection Service (APHIS)–funded research at Kansas State University, Purdue University, and Texas A&M. APHIS contracted with the three institutions to study issues on massive carcass disposal and make recommendations for APHIS. The results would later be considered for potential revisions to the field manual for first responders and to identify further needs for research. The authors led the "Local, State, and Federal Regulation: Interagency Relations" section of the project. The following provides a look into the findings involving a roundtable discussion of a hypothetical agroterrorism incident (Borron and Cain 2004).

Background Information on Agency Involvement in Emergency Response

The history of massive animal carcass disposal disasters in the United States and other countries indicates many interagency issues and possible subdisasters for those agencies if steps are not taken ahead of time to anticipate problems. For example, the foot-and-mouth disease outbreak in Great Britain in 2001 showed how a lack of cooperation between jurisdictions and local and national agencies resulted in:

- Extended disease control issues
- A loss of human lives (suicides)
- A catalyst for change of a national agency (The United Kingdom's Ministry of Agriculture, Fisheries and Food [MAFF] became the Department for Environment Food and Rural Affairs [DEFRA].)

While not all potential interagency problems can be anticipated and addressed in advance, two of the actions that might prevent a disaster from taking larger tolls are education and facilitation. Factors related to education include:

- Better understanding of the National Incident Management System (NIMS) by agricultural industry leaders and participants. Note: In 2004, NIMS replaced the Incident Command System (ICS), which is a term still used by many emergency managers and is still taught because not all state and county trainings have evolved to use NIMS. As NIMS becomes better understood it will enhance interagency relations. The NIMS movement will use the same basic concepts as ICS. NIMS uses multiagency oversight that President George W. Bush provided with the unified Department of Homeland Security.

- Better understanding of the ICS/NIMS, standard operations procedures, and agriculture by county governments and agricultural groups.
- Better understanding of agriculture by the emergency management and county government systems.
- Better understanding of agricultural disaster response by state and local agencies (public health, legal, etc.).

A primary factor related to facilitation is encouragement of periodic (annual or semiannual) meetings at the state level to discuss specific operational, legal, and future research needs in the area of animal disaster management.

Overview of the Problem

When a disaster strikes, a number of agencies respond, depending on the type and magnitude of disaster. When multiple-agency involvement becomes a factor, the efficiency of interagency relations and communications is important. Such coordination is a key component of a successful outcome. The questions what works, how does it work, and what should be implemented are important when examining ways to strengthen the existing infrastructure of agencies that respond to state disasters. In the event of a major disaster, proactive interagency coordination will aid in the response efforts, whereas the lack of coordination will hinder the progress of necessary actions. Specifically, steps taken within agencies to provide SOPs that enhance an agency's response, as well as interagency response, are critical to successful outcomes.

Test Exercise: Indiana Example

In December 2001, the Indiana State Emergency Management Agency (SEMA) put into effect a revised version of the comprehensive emergency management plan (CEMP). The CEMP is a checklist requiring all state agencies to develop and implement SOPs and standard operating guides (SOGs). The function of CEMP is to outline expected protocols for disasters most likely to affect Indiana, designate the primary coordinating agency for a given disaster, and determine the supporting role of other agencies (Indiana State Emergency Management Agency 2003).

In Indiana, two actions will enhance the response efforts during a major agricultural disaster. First, acting agencies need to know they are part of the CEMP plan. Second, more people within agencies should have a comprehensive awareness and understanding of all others involved, in addition to understanding their own agency's SOPs. In order to enhance the functionality of the CEMP, SEMA also incorporates the use of the Incident Command System (ICS) during the management of a disaster. As noted earlier, this will evolve into NIMS.

The ICS is a standardized response management system. As an "all hazard, all risk" approach to managing crisis response operations as well as noncrisis events, this system is organizationally flexible and capable of expanding and contracting to accommodate responses or events of varying size or complexity (United States Coast Guard 1998).

The ICS has four functional areas:

- Operations: This area includes all activities directed toward reducing the immediate hazard, controlling the situation, and restoring normal operations.
- Planning: This area includes the collection, evaluation, dissemination, and use of information relative to the development of the incident and the status of resources, as well as creation of an action plan.
- Logistics: This area provides all support needs; orders all resources from off-incident locations; and provides facilities, transportation, supplies, equipment maintenance, meals, communications, and medical services.
- Finance: This area tracks all incident costs and evaluates the financial considerations of the incident.

In order to pull all elements of disaster management together, SEMA takes a top-down approach. A general response plan is developed for disasters most likely to take place in Indiana. For each plan, a number of specific disaster situations are addressed. To deal with these particulars, annexes are created. Certain instances require the elaboration of annexes or the narrowing of specific responsibilities to agencies or organizations. At this point, an SOP is created for more finite guidance to the annex. Overall, the ICS provides a flexible structure to deal with changing disaster scenarios and the various annexes/SOPs that apply.

Text Exercise: Methods and Process

In order to consider the current status of interagency coordination in Indiana, the authors took the initial step in 2003 to design a high-magnitude disaster on paper (Appendix A) that would demand the involvement of a number of agencies from a variety of areas. The scenario used in this project was called Dead Animal Disease (DAD). The intention was to create a situation that placed the participants at a specific point—two weeks into an unknown animal disease with an anticipation of a massive carcass disposal—that would present a number of unanswered questions.

The second step was to organize a roundtable discussion that would provide the agencies with an opportunity to come together as a group and discuss the expected roles and responsibilities of each agency during the hypothetical disaster. The following agencies participated in the project:

- County-level board of health
- Indiana Board of Animal Health (BOAH)
- Indiana Counter-Terrorism and Security Council (C-TASC)
- Indiana Department of Environmental Management
- Indiana Department of Natural Resources
- Indiana Office of the Commissioner of Agriculture
- Indiana State Chemist Office
- Indiana State Department of Health
- Indiana State Emergency Management Agency
- Indiana Public Health Association
- Purdue Animal Disease Diagnostic Lab
- Purdue University Cooperative Extension Service
- U.S. Attorney General's Office
- USDA's Farm Service Agency

All participants were provided the scenario in advance. In addition, they were asked to answer a list of questions (Appendix B) regarding their roles and actions for the CEMP at two weeks into the disaster. These answers were collected, organized into one document, and mailed to everyone for their review prior to the discussion.

The individuals who participated in the discussion were directors from various areas of their respective agencies, including administration, communications, and operations. All participants were chosen based on the leadership role they would play the moment their agency became involved in the response efforts.

At the onset of the roundtable discussion, individuals were allowed the opportunity to share additional information in regard to their previous responses. At this point, many questions were raised as to who would be responsible for what and how it would be accomplished. As the discussion continued, the group was asked to consider the areas of cooperation among responding agencies, as well as future actions that should be considered in order to improve interagency coordination. All participants provided valuable information in regard to their agencies' roles and responsibilities during the course of the hypothetical animal disaster. Much information was provided for consideration, identification, and in some cases, realization for the first time by others involved. For the most part, concentration fell on three main areas: response, communication, and education. The following are a number of comments and questions that were discussed as a group:

- Response
    - While BOAH and SEMA know who is in charge, do a critical number of other agencies know who is in charge?
    - Who should formulate and make a public announcement at the appropriate time?
    - What is the level of public health significance of an agricultural event?
    - What audiences are affected? Affected individuals have a right to know what is taking place, and in the event of quarantine, people will demand freedom of movement and commerce.
    - Would initial actions and decisions be committee-based?
    - How will staffing needs be fully met?
    - At what time is it appropriate for an agency to begin responding?
    - Should the subject matter expert and the jurisdictional authority be the same person?
    - What are the legal and jurisdictional issues? What do you legally have the right to do?
    - SEMA will prepare and distribute situational reports of other agencies as a way of sharing information.
    - Planning for too narrow of a perspective puts preplanning resources in the wrong place. Having a specific plan for every incident is impossible; sometimes what you have has to be enough.
    - Perhaps the memorandums of agreement (MOAs) take precedence; taking precedence overall is the continuity of government to show the agreement of function and cooperation.

— Considering the cooperative agreements as well as identifying possible cooperative research that exists—in many ways, this is already being done with carcass disposal in regard to land layout and site identification.

- Communication
  — Animals and animal by-products leaving Indiana will be considered tainted. We must communicate to the public the real health risks and actions taking place to remove the risks and restore a healthy food supply.
  — Communication is the key factor throughout the entire situation—a communication center has to be up and ready, first and foremost.
  — When something is unknown (e.g., DAD), offering a timeline for identification could be nearly impossible.
  — The sharing of information from one level to the next should be kept consistent among multiple agencies.

- Education
  — Appropriate agencies with proven records should be utilized for public education.
  — Educational efforts are key to the cooperation of the affected public during necessary response efforts. Examples include educating people who could be inhibitors to the eradication of the disaster at hand, informing people of the possible threats created by moving animals, and educating people on the safety of the environment around infected areas/farms (i.e., water/fish from nearby stream).
  — Educating more leaders/figures who need to be key players in developing plans and communications.
  — The Food and Drug Administration (FDA) and USDA's Extension are in prime positions to serve as resources of information and education.
  — Every county should have emergency response training in place.
  — All agencies can learn from past events. Examples mentioned were the 2003 outbreak of *Ralstonia solanacearium*, race 3 biovar 2 in geraniums; the 2003 outbreak of monkey pox in prairie dogs; and the 2001 outbreak of foot-and-mouth disease (FMD) in Great Britain's livestock. In the *Ralstonia* situation, the USDA needed a quicker confirmation and action plan that was communicated clearly to all involved agencies. In the monkey pox situation, the communications from the Department of Health were not activated quickly enough because the department assumed it was

not human health–related, and the FMD issues were explored at the beginning of this document. All three situations provided insights and learning opportunities as to how agencies would act (or not act) at the finding of an outbreak.

- Recommendations
  - Strengthened cooperation is needed not only between government agencies but also with industry and the organizations representing the public.
  - Take advantage of resources available for use where needed in the response to a disaster (i.e., superior Food and Drug Administration and Environmental Protection Agency labs).
  - Providing reassurance to all those concerned could mean taking actions that are not necessary for the event but are necessary for public easement. Actions that deal with perceived issues as well as real issues and communicating that message are necessary to reassure the public.

Test Exercise: Strategies to Deal with Issues

The hypothetical event (DAD) was directly animal-related, which automatically placed BOAH as the lead agency. However, as events unfolded, other areas of expertise were in demand. The testing capabilities of the Purdue Animal Disease Diagnostic Lab (ADDL) were required because the agent causing the animals' sickness and subsequent death was unknown. In addition, the resources of contracted companies and agencies, such as the Indiana Department of Environmental Management (IDEM), were needed because approximately 37,000 animal carcasses required handling and disposal.

Often times, certain assistance is necessary due to events that take place indirectly to the overall disaster. The Indiana State Department of Health (ISDH) should have been called upon for three initial reasons:

- The agent/disease was unknown, raising the question of whether or not it was zoonotic, which presents the consideration of how it could affect humans.
- A massive carcass disposal issue was ensuing, which inevitably creates a human health and safety issue.
- Such a large disaster would find its way to the media outlets, causing a possible public perception of fear and concern about such things as the food and water supply. (Note: As identified in past exercises, additional agencies are brought into the "mix" at the request of the lead state agency or at the recommendation of SEMA.)

After examining the collected information and considering the open-ended questions posed to agencies at the two-week point in the animal disaster scenario, the next step was to consider what currently works in the state of Indiana. Relationships between agencies with well-defined responsibilities work well during a disaster. For instance, in the case of a known animal disease outbreak, BOAH and SEMA establish a teamed response with the necessary chain-of-command organization quickly in place through the common practice of the ICS.

In the instance of the animal disaster scenario used in this project, BOAH and SEMA would be the initial organizers. As the events of a disaster continue to unfold, more responding agencies are required to become an integral part of the ICS. However, some key agencies may not have a good understanding of how this system functions. As a result, the organization of the four functioning ICS areas (operations, planning, logistics, and finance) potentially could be slowed.

State agencies are part of the emergency response system, but those at the local level are involved as well. In the DAD disaster scenario, the incident was contained within a 25-mile radius of the Indianapolis airport. As a result, county law enforcement and emergency personnel were involved from the beginning and/or as events unfolded. Such involvement demonstrated an overlapping of MOAs of the local or county agencies with the functionality of the CEMP at the state level. This will result in local action versus state action. For example, as the number of dead animals rises, carcass disposal issues will need to be addressed, which would result in possible local jurisdictional conflicts and authority issues between county and state. In addition, county governments may not have a good understanding of ICS and agriculture's specific needs.

Test Exercise: Outcomes

An idealistic approach to a disaster would be to know in detail what needs to be done, what protocols need to be enacted, and who is going to take the lead. However, no real-life disaster plays out as a textbook example. General disaster plans are created with a number of annexes and SOPs attributed to specific situations. Several areas should be addressed to achieve a higher level of preparedness and response, regardless of the tragedy or the number of agencies involved:

- An interagency working group should be created that meets twice a year and consists of at least the state environmental, animal health, public health, contract service, emergency management, extension service, transportation, and wildlife agencies.
- An analysis should be conducted of the agencies' (state and county) awareness level of the functionality of the CEMP and its components,

as well as the overall functions of the ICS. Have enough agencies been included? Are there enough training opportunities for agency employees? Do the involved agencies have a well-established representation of their SOPs within the annexes of the CEMP?

- A training program that:
  - — Requires ICS training for all agencies involved in the CEMP—state and county level. The training should include enough people from various agencies to ensure a widespread understanding of the ICS and various agencies roles.
  - — Establishes programs at the county level to bridge the gap between the legal system and agricultural issues in a biosecurity event.

Test Exercise: Reflections

The assessment of interagency communication began with an attempt to consider the relationships that should exist across platforms for the most effective response to a high-magnitude disaster. Therefore, the creation of a situational disaster requiring agencies to approach the problem from opposite directions was necessary. A list of potential participants was created through examination of possible required resources. However, as was expected, missing entities were not identified until the roundtable discussion took place. In hindsight, valuable information from individuals at the local and federal levels was lacking.

Once information from all participants was collected and organized, it became evident that the problem might not exist entirely with interagency communications but, rather, with the total understanding of the incident command system. Therefore, a stronger emphasis was placed on training rather than communication during the development of possible solutions. If this project were repeated, the focal point would move from the quality of communication taking place between agencies during a disaster to the comprehensive training provided within agencies on how the ICS needs to function to be successful. Necessary communication will begin to improve if all involved individuals and their respective agencies are fully aware of how their role will develop in a disaster. Once that is established, areas still lacking in interagency communication should be addressed.

Test Exercise: Critical Research Needs

This study demonstrated that more could be known about how key agencies will react to a massive animal carcass disposal situation. While facilitation of this process will help agencies discuss their respective issues, some issues will not be addressed by agencies due to prioritization and current workloads. In other words, many agency professionals will not feel the need to put a high priority on animal carcass disposal issues. They will not be inclined to dedicate staff time to a theoretical issue when they have enough real issues to deal with at the present.

Research into (and summarization of) the laws, regulations, guidelines, and standard operating procedures of key state agencies involved in responding to catastrophic carcass disposal events is needed.

▶▶ Conclusions

In conjunction with the USDA-funded project, a roundtable discussion was organized within the state of Indiana to provide an opportunity for representatives from state agencies involved in responding to a foreign animal disease outbreak to come together to discuss the expected roles and responsibilities of each agency during a hypothetical disaster. Results of this roundtable discussion demonstrated that

- More could be known about how critically involved agencies will react to a specific agricultural emergency situation.
- In an environment of short-staffing and high workloads, agency personnel will likely not place a high priority on planning for theoretical agricultural emergencies.

Therefore, to facilitate planning efforts and provide structure for interagency discussions and exercises, research into (and summarization of) the actual laws, regulations, guidelines, and SOPs of key agencies is warranted on a state-by-state basis. This research is critical to the development of comprehensive plans for state and county governments to more easily identify their roles. These could be used in training programs for state and local agencies to develop pertinent standard operating procedures and memorandums of agreement.

References

Borron, A., and S. Cain. 2004. Regulatory Issues and Cooperation. In *Carcass Disposal: A Comprehensive Review*. Kansas State University: National Agricultural Biosecurity Center.

Bush, G. W. 2004. Homeland Security Presidential Directive/HSPD-9. White House. http://www.whitehouse.gov/news/releases/2004/02/20040203-2.html (accessed March 17, 2005).

Committee on Biological Threats to Agricultural Plants and Animals, National Research Council. 2002. *Countering Agricultural Bioterrorism*. Washington D.C.: The National Academies Press.

Dyckman, L. 2003. Bioterrorism: A Threat to Agriculture and the Food Supply. United States General Accounting Office. http://www.gao.gov/new.items/d04259t.pdf (accessed March 17, 2005).

Indiana State Emergency Management Agency. 2003. Indiana Comprehensive Emergency Management Plan. Indiana State Emergency Management Agency. http://www.in.gov/sema/emerg_mgt/cemp.pdf (accessed March 18, 2005).

Kaiser Permanente Health Care Continuity Management Team. 2002. Terrorism Preparedness: Ensuring Successful Collaboration Between Government Agencies and the Health Care Industry to Prevent and Respond to Acts of Terrorism. Kaiser Permanente. http://www.vdh.state.va.us/bt/Secure%20VA.%20Subpanel %20Meeting%2010.22.02/Terrorism%20Preparedness%20White%20Paper%20% 20091502.doc (accessed March 3, 2005).

Monke, J. 2004. Agroterrorism: Threats and Preparedness. CRS Report for Congress. Congressional Research Service. Library of Congress. http://www.fas.org/irp/ crs/RL32521.pdf (accessed March 2, 2005).

National Animal Health Emergency Management System. 2003. Animal Health Emergency Management Activities State Reports: 2002–2003 Successes and Challenges." National Animal Health Emergency Management System. http://www.usaha.org/NAHEMS/n0203sts.pdf (accessed March 4, 2005).

Parker, Henry S. 2002. Agricultural Bioterrorism: A Federal Strategy to Meet the Threat. McNair Paper 65. Institute for National Strategic Studies, National Defense University. http://www.ciaonet.org/wps/pah01/pah01.pdf (accessed March 4, 2005).

RAND National Defense Research Institute. 2003. Agroterrorism: What Is the Threat and What Can Be Done About It? RAND Corporation. http://rand.org/ publications/RB/RB7565/RB7565.pdf (accessed March 17, 2005).

United States Coast Guard. 1998. Incident Command System. United States Coast Guard. http://www.uscg.mil/hq/g-m/mor/Articles/ICS.htm (accessed March 18, 2005).

United States Government Accountability Office. 2005. GAO-05-214 Homeland Security: Much Is Being Done to Protect Agriculture from a Terrorist Attack, but Important Challenges Remain. United States Government Accountability Office. http://www.gao.gov/htext/d05214.html (accessed March 17, 2005).

Appendix A

Indiana Biosecurity and Public Health Roundtable, Situational Setup

▶▶ What:

A breakout of Dead Animal Disease (DAD)—This is an unknown disease. At two weeks into the disaster, affected animals include cows, pigs, and chickens. Symptoms include internal bleeding and massive respiratory problems. The incubation period appears to be five to seven days with death occurring three days later. The spread appears to be rapid. Confidence is high that it does not affect humans, but such a concern is not 100 percent ruled out.

▶▶ Where:

A total of seven farms within a 25-mile radius of the Indianapolis Airport are reporting the disease.

▶▶ When:

The first reports are from a dairy farm (A) and poultry farm (B) on July 16. The other five farms report symptoms four days later on July 20.

▶▶ Details:

Farm A: 1,000 dairy cows
Farm B: 16,000 chickens
Farm C: 12,000 swine
Farm D: 500 beef cattle
Farm E: 5,000 swine
Farm F: 1,500 dairy cows
Farm G: 1,000 beef cattle
Total Number of Animals: 37,000

By July 17, an unknown disease, which is being referred to as DAD, is identified within the confines of farms A and B; 675 cows and 7,350 chickens are showing symptoms for the mysterious disease. Two days before the confirmation

(July 15) a feed truck had made rounds to these two farms, as well as ten others. By July 20, five of the ten are reporting symptoms. On the same day, 10 percent of the infected animals on farms A and B have died. On July 21, the truck is quarantined. On July 18, concerned neighbors near farms A and B report to the Dawson County Sheriff that a white sedan was seen near the farms' premises. Both accounts verify that the sedan had rental plates and was carrying three or four people. To date, there is no evidence of this vehicle, or others, being on all seven farms.

▶▶ All Farms Ship to Markets:

- Farms A (milk daily), D (2x/yr), E (1x/week), G (2x/yr)—state shipping
- Farms C (1x/week), F (milk daily)—interstate shipping
- Farm B (eggs daily)—international shipping

▶▶ Those Affected:

- The infected farms are experiencing catastrophic losses. At minimum, 37,000 animals will have to be dealt with for mass carcass disposal.
- Surrounding land and uninfected farms that are located in the established quarantined perimeter (a three-mile radius) around each infected farm. Such quarantine would institute a complete halt to all business which concerns movement outside of the property.
- People/Public could be affected in four ways:
  — Those in quarantined zone could be deemed immobile for an enforced amount of time.
  — Massive carcass disposal issue = public health issue.
  — Public perception and concerns—a poor understanding of DAD and a fear of the safety of associated animal products bought from grocery store shelves or supplied to school lunch programs.
    – One problem is that DAD is so closely timed with SARS. Some feel strongly it could affect humans. Therefore, the public fear level is increased.
  —Still has not been ruled out that this disease is not zoonotic.
- The national dairy, pork, poultry, and beef markets experience a devastating drop in prices and trade capabilities.
- Already, scores of national reporters are camped out on the west side of Indianapolis and are demanding information.

▶▶ Questions/Assumptions/Scenario Changes:

- Can all shipped meat, milk, eggs, and live animals from infected farms be tracked?

- Characteristics of this disease: What is the rate of spread? How long is the incubation period? What are the potential vectors? Can it be spread by contact, air or other animals?

- What are the appropriate biosecurity procedures that the animal care specialists must take to safeguard themselves and unaffected animals?

- Will other species, such as wildlife, have to be examined or destroyed because of this outbreak? If so, how will this hinder personnel and the logistics of controlling the situation?

- Possible assumption: DAD is a genetically modified organism.

- Possible assumption: It is suggested that the disease was spread into confinement buildings through an aerosol sprayed into the air intake. This makes the disease deadly at those operations. But, because of modern confinement and current biosecurity habits, the disease does not seem to be spreading as fast as it could.

- Scenario change: The county sheriff, in cooperation with a local citizen, finds a suspect container with trace amounts of an unknown substance that is currently being investigated. This container was found in a ditch just outside the city limits of the Dawson County Seat.

Appendix B

Indiana Biosecurity and Public Health Roundtable, Questions Posed to Participants

The accompanying Dead Animal Disease (DAD) scenario explains a hypothetical outbreak of an unidentified disease that is suspected to be genetically altered. Please refer to this scenario as you answer the following questions (if your answers require more space, please use the back of this page or attach additional pages, if necessary):

1. The state of Indiana has a Comprehensive Emergency Management Plan. Is your agency represented in that plan? ___ Yes ___ No ___ Don't know.

2. This plan calls for standard operation procedures (SOPs) with guides and plans to support it. In reference to the "DAD Scenario," does your agency have SOPs that apply? ___ Yes ___ No ___ Don't know.

3. Considering those SOPs and the "DAD scenario," at two weeks into the disaster:

   a. What protocols would have been completed by your agency?

   b. What continuing steps would you expect your agency to take?

4. For the DAD disaster, what Memorandums of Agreement (MOAs) or Memorandums of Understanding (MOUs) do you think are already in place to aid in interactions with other agencies?

5. For the DAD disaster, what MOAs or MOUs do you think need to be in place in the future to aid in interactions with other agencies?

6. What problems do you feel will surface if a disaster of this nature and magnitude appears in Indiana?

Chapter 3

Pedagogical E-Learning Framework Using Advanced 3D Visualization for Bioterror Crises Communication Training

KRISHNA P. C. MADHAVAN, MOHAN J. DUTTA-BERGMAN, AND LAURA L. ARNS
Purdue University

▶▶ Introduction

The events of 9/11 and the subsequent anthrax attacks have triggered a new era of national security concerns that require first responders to be highly trained and efficient in the use of technology, communication skills, and crisis handling. In particular, the threat of bioterrorism is real and requires a constant state of preparedness at multiple levels of national, regional, and state emergency organizations. In an effort to enhance bioterror response readiness and to analyze systemic weaknesses thereof, the U.S. federal government in collaboration with the Advancing National Strategies and Enabling Results (ANSER) Institute for Homeland Security, the Center for Strategic and International Studies, the Johns Hopkins Center for Civilian Biodefense Studies, and the Oklahoma City National Memorial Institute for the Prevention of Terrorism conducted the *Dark Winter* simulation in the winter of 2002. This exercise simulated how emergency agencies and their personnel reacted to a biological weapons attack on the United States. The post hoc analyses, highlighting the enhanced need for bioterror crises communication training, concluded that "dealing with the media will be a major, immediate challenge for levels of government. Information management and communication (e.g., dealing with the press effectively, communication with citizens, maintaining the information flows necessary for command and control at all institutional levels) will be a critical element in [bio-terror] crises/consequence management" (ANSER Institute for Homeland Security 2003). The complexity and span of bioterror crises communication is such that no single field of scientific enquiry can possibly take on the responsibility for developing pedagogical paradigms for training personnel responding to bioterror events. It falls, therefore, upon an interdisciplinary group of experts to develop new and innovative strategies that improve training for professionals in the field of healthcare and public health, with a particular focus on bioterror crises communication. Such training must be an essential component in the health sciences, as well as in communication programs at schools and universities.

The ability of healthcare and communication professionals to successfully respond to future bioterror incidents will rely heavily on (1) systematic use of advanced technologies in handling the tremendous influx of data triggered by the bioterror incident, (2) significant knowledge of communication technologies that will enable real-time decision making, (3) keen awareness of important theories in crises communication response, and (4) significant hands-on training in real-time bioterror communication handling techniques. The focus of this chapter is to outline a framework that combines extant global e-learning specifications and 3D visualization tools and techniques within the broad tenets of crises communication theories. The next section presents this pedagogical framework.

▶▶ Core of the Proposed Framework

Whereas bioterror crises communication has serious implications for all parties involved in a particular scenario, in pedagogical terms it represents a problem-solving exercise of the highest order. Therefore, the basic pedagogical principles that apply to the promotion of problem-solving skills among students are also applicable in the case of bioterror crises communication training. In the context of bioterror crises communication, the training scenarios need to go well beyond providing students with an assortment of knowledge sources to draw upon in real-world scenarios, and instead facilitate deep learning. The extant theoretical perspective available from the field of science education suggests that emphasis on core concepts has more value than just presenting knowledge sources—namely, facts (Chi et al. 1981; Glaser

1992; Schneider et al. 1993). Currently, there are no simple mechanisms to incorporate real-world crises communication scenarios into regular training for healthcare and communication professionals. The costs and logistics involved in organizing full-fledged training scenarios are prohibitive at best. This leads to scenarios where trainees are provided with a number of knowledge resources but few opportunities to internalize this knowledge. As a result, crises communication trainees may experience an overflow of knowledge with no avenue to actually synthesize it. "The ability to capture knowledge such that it can be analyzed, reused, and shared with others, thus developing a spiral of more new knowledge creation, is perhaps the most powerful promise information technology can provide. The impact on learning of just-right information flowing to the right place, person and time, cannot be overstated" (Hodgins 2000). The advent of learning objects marks the arrival of one such IT innovation that could potentially allow for the better capture and dissemination of knowledge. The definition of learning objects has been prescribed by the International Electrical and Electronics Engineers' (IEEE) Association.

> Learning objects are defined [here] as any entity, digital or non-digital, which can be used, re-used or referenced during technology supported learning. Examples of technology supported learning include computer-based training systems, interactive learning environments, intelligent computer-aided instruction systems, distance learning systems, and collaborative learning environments. Examples of learning objects include multimedia content, instructional content, learning objectives, instructional software and software tools, and persons, organizations, or events referenced during technology supported learning. (IEEE LTSC 2004)

Learning objects are an artifact of the standardization processes that are currently seen across the IT industry. Standards-based learning technologies—such as learning objects—play a critical role in facilitating cross-disciplinary collaboration and community building (Wiley 2003) by promoting interoperability and reusability. While interoperability refers to the ability of learning content to function across a variety of software environments, reusability highlights the ability of the learning content to morph itself to suit the context of use (Advanced Distributed Learning 2003). Learning objects are digital entities that can have varying granularity—from simple video clips to advanced immersion-based simulations such as the ones that are required to deliver bioterror crises communication training. Learning objects are, by their very definition, required to include measurable objectives—therefore, highly suitable for including assessment components (Madhavan 2004a). Furthermore, learning objects are perfect delivery vehicles for work created by an interdisciplinary team of experts (Madhavan 2004b; Acker et al. 2003) such as being proposed through this framework.

At the core of this methodology is the use of appropriate metadata. The 3D visualization-based simulations that are created need to be enveloped with metadata that describe aspects of the simulations—such as time required to complete the simulation, appropriate levels of use (namely undergraduate, graduate, or professionals currently in the field), and also information to ensure that students can be tracked and assessed consistently. By virtue of being associated with the relevant metadata, the format for which is very strictly defined by IMS (note: there is no expansion for this organization; it is simply referred to as IMS [see IMS 2001]), the 3D visualization-based simulations become transformed into global e-learning specification compliant learning objects.

The important question that needs to be addressed at this stage is: Why is this approach of blending global e-learning specifications with 3D visualization an effective strategy for developing crises communication training? The next section addresses this question.

▶▶ Merging Virtual Reality and Global E-Learning in Bioterror Crises Communication Training

The concept of virtual reality (VR) was first formally defined in 1965 when Ivan Sutherland introduced his Ultimate Display (Sutherland 1965). Since that time, the term *virtual reality* has been used in a wide variety of contexts. Unfortunately the term is often misused by the popular media in reference to such activities as 3D movies, video games, or the World Wide Web. In the work presented here, VR refers to "immersive, interactive, multi-sensory, viewer-centered, three-dimensional, computer-generated environments and the combination of technologies required to build these environments" (Cruz-Neira 1995). *Immersive* means that users have the illusion of being a part of the environment; they are not just observers on the outside looking into the computer-generated world. *Interactive* means that users are able to affect the state of, and receive feedback from, the virtual world.

The main component associated with virtual reality is computer graphics, and the devices used to present them. Many people associate 3D vision with virtual reality. This is the technique known as *stereoscopic graphics*, often referred to as *stereo*. The idea behind computer simulated stereo vision is to emulate natural human vision, in which both eyes see slightly different views of the world due to their separation. The brain is able to fuse these two views into a single image that contains information about the depth of the object(s). This process is known as *stereopsis* (Burdea and Coifett 1994). A variety of methods exist for emulating natural stereo vision for users of a virtual environment, such as liquid crystal display (LCD) shutter glasses or polarized glasses.

A variety of graphics displays are available for use in a virtual reality system, including head-mounted displays (HMDs), stereo workstation displays (so-called fishtank VR), and projection-based systems in which graphics are projected onto display screens. For applications in which collaboration with other users is vital, small displays such as HMDs may be inappropriate because they present graphics for only a single user. Large projection-based systems such as the CAVE (Cruz-Neira et al. 1993) offer the ability for several participants to share and discuss displayed information. Recent developments such as the Fakespace FLEX provide for the possibility of additional participants and a reconfigurable working space. Other solutions such as tiled walls create the ability to display information at a much higher resolution than with a single projection display. Studies such as Arns et al. (1999) demonstrate that stereoscopic projection displays outperform traditional workstation displays for providing users with the ability to understand and analyze certain types of data.

The purpose of this section is to demonstrate that 3D simulations are indeed a viable, well-established pedagogical practice. While the use of simulations has a long history in the areas of technology and instructional design, the simulations have mostly been 2-dimensional and minimally interactive. Furthermore, the ability to tie in assessment components to the actual simulation environment has not even evolved fully, let alone been fully exploited. By using the term *minimally interactive*, the intention is to point out the relatively low levels of interactivity of 2D or Web-based simulations in comparison to the 3D immersive simulations that utilize advanced data perceptualization methods. Simulations in an educational context are defined as "educational exercises that provide students the opportunity to 'role play' the concern of the stakeholders" (United States Institute of Peace n.d.). There is compelling evidence in the literature on teaching and learning that shows that student learning is facilitated to a great extent by the use of educational simulations. Billhardt (2004) points out that "simulations offer huge advantages over lectures, handbooks, or on-site trainers. They engage students while helping them retain and apply what they learned. They allow people with a wide range of learning styles to meet learning objectives at their own pace." The ability of educational simulations to allow learners to interact with learning materials at their own pace allows the selection of appropriate conceptual areas that the learners find problematic. The fundamental assumption behind using educational simulations is that when students apply the knowledge that is conveyed in the form of theories and abstract concepts to solve practical problems, *deep* learning occurs. The responsibility for the learning process is now partially shifted into the hands of the learner (de Jong et al. 1998). The learner, therefore, is transformed from a simple recipient of information to active collaborator in the teaching and learning process, and the teacher is moved to the role of a facilitator (Oliver et al. n.d.).

Simulations have been used repeatedly to convey key concepts in science, technology, engineering, and mathematics (STEM) education. The evidence of success and the various limitations resulting through the use of simulations in the STEM areas are documented (Mackenzie et al. 2001; Jayakumar et al. 1994; Vieth et al. 1998; and Eylon et al. 1996). Through the analyses of the above studies, it is clear that simulations should recreate the skill-use environment closely by adhering to the rich stimuli that exist in the real world (Brown et al. 1989; Young and McNeese, 1993). To create simulations that can allow the relatively complete mobility of real-world characteristics to the simulation environment requires a tremendous amount of resources. In the case of computer-based simulations of extremely complex environments, the computational cycles require reconstructing the real-world entities that are not readily available in the traditional classroom setting. Areas such as nanosciences utilize large computational clusters to provide students with the necessary power to complete simulations.

In the case of bioterror crises communication simulations, there are a large number of theoretical and practical constraints that need to be reconstructed by the instructors to convey appropriate training. The number of data points and variables handled by the simulated environment require full-fledged simulations that have extremely easy access to computational resources. This, perhaps, explains why materials for training students on bioterror communication situations have not been produced. The fundamental assumption behind the methodology proposed in this chapter is that modern-day problems require computational resources far beyond what is available in a traditional classroom. Web-based simulations that are currently available on the Internet do not approach the level of computational resources that are required for bioterror crises communication training simulations. It is in this context that environments that are specifically geared to construct advanced 3D visualization-based simulations need to be employed.

The use of 3D visualization-based simulations adds significantly to the authenticity of the learning experience. This blended approach to crises communication training integrates emerging technologies into a perceptually coherent strategy for the construction of instructional materials. The use of this framework, called perceptualization to stress that the integrated whole of all human experiences is bigger than the sum of individual parts, not only fashions new paradigms for the field of crises communication but also closes infrastructure gaps to exploit emerging technology in data perceptualization for educational purposes, thus embedding the learners in authentic contexts (Rogoff 1984; McLellan 1994; and Spiro et al. 1991 for a discussion on the nature and advantages of using authentic contexts for teaching and learning).

The application of advanced computational techniques to generate 3D visualization-based simulations along with all related metadata provides avenues that researchers can use to study various processes, constructs, and theoretical notions that were not possible before. The ability offered by visualization and virtual environments to control and observe every aspect of the learning process will allow the definition of research questions that offer new insights to all collaborating areas—particularly to the overall betterment and advancement of the teaching and learning process. The use of such types of advanced computational techniques for bioterror crises communication training materials development is of particular importance given the need for near-reality or authentic learning environments in the training process.

There have been some arguments presented to highlight the need for simplifying the complexity of the content presented at the introductory levels with increasing complexity introduced at more advanced levels (Sandberg and Wielinger 1992; Riegeluth and Schwartz 1989). This approach to instructional design is consistent with the systems model (Dick 1991; Dick and Carey 1990; Gagne et al. 1992). One of the significant areas of concerns with bioterror crises communication training is to resist the temptation to oversimplify the content delivered to students. A direct criticism of this tendency to offset complexity of content by oversimplifying classroom materials is apparent in the following statement: "Simplification of complex subject matter makes it easier for teachers to teach, for students to take notes and prepare for their tests, for test-givers to construct and grade tests, and for authors to write texts. This results in a massive 'conspiracy of convenience'" (Spiro et al. 1987).

The next section of this paper will discuss theoretical and practical factors that are needed for developing appropriate materials for bioterror crises communication training.

▶▶ Bioterror and Crises Communication Training

Crisis is defined as a "specific, unexpected, and non-routine event or series of events that create high levels of uncertainty and threaten or are perceived to threaten" the community and its citizens (Seeger et al. 1998). One of the critical features of a crisis is the high level of uncertainty associated with it (Coombs 1999; Babrow and Dutta-Bergman 2003; Noll 2003). In many instances, the key stakeholders in the crisis do not really have a clear idea about the nature of the crisis, its origins, and ways to deal with it and prevent it (Babrow and Dutta-Bergman 2003; Sellnow and Seeger 2001). The high level of uncertainty brought about by the crisis, accompanied by the set of novel conditions the crisis creates, generates a mixture of public responses such as confusion and anxiety, stress and fear about the future, actions that lead to additional chaos, and inaction in situations that need public action. This is accompanied by intense media scrutiny and coverage of the event. Typically, the crisis threatens the credibility of the involved organizations in the public arena, calling for the development of well-planned communication strategies (Sellnow and Seeger 2001).

In recent years, crisis management research has acknowledged the critical role of communication (Fishman 1999; Sellnow and Seeger 2001). Communication is not only critical in the domain of postcrisis response strategies but is a continuous process through the life of the organization (Coombs 1999; Sellnow and Seeger 2001). It offers the conduit through which intraorganizational and interorganizational communication are managed and maintained in pre- and postcrises environments. Extant research in communication points out five key areas of crisis communicative strategy for effective message development in the realm of crises: timing, content, channel, appeal, and source. The goal of the visualization-based pedagogy is to introduce the students to best practices in the realm of these key strategic areas through the exercise of active selection in response to crises. Figure 1 explains the key ingredients of crisis communication strategy.

Timing Strategy
How should the crisis response strategy be timed? A significant portion of existing research deals with the timing of crises responses. For instance, in response to allegations that the Audi 5000 was prone to sudden acceleration at random instances, culminating in the *60 Minutes* news coverage, the organization

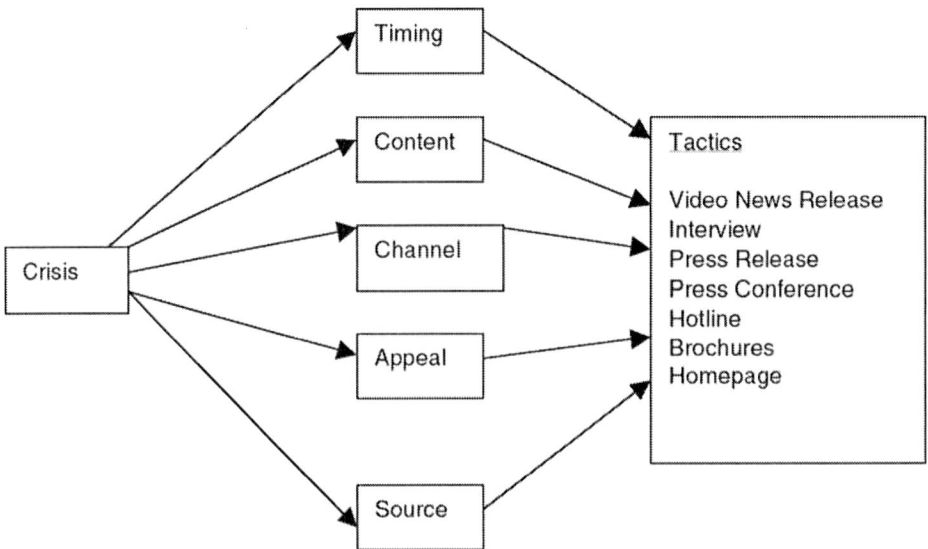

Figure 1
Model depicting the key elements of crisis response training

decided to wait it out. Scholars studying the crisis pointed out that the reputation of the organization was damaged because it did not respond immediately to the crisis. The crisis response to the floods of the Red River Valley, however, was effective because it was timed for immediate response; relevant organizations responded immediately and communicated clearly about the crisis. Much of the published communication research suggests that an early and open organizational response to the crisis minimizes the damage brought about by the crisis to the organization's credibility (Fishman 1999; Seeger et al. 1998; Seeger et al. 2001).

Timing is particularly relevant for bioterror crises because organizations need to communicate response strategies with the public in a manner that responds to the immediate nature of the event and yet does not create panic in the population (Seeger et al. 1998; Seeger et al. 2001). For instance, research on the diffusion of news of the terrorist attacks of 9/11 demonstrated that the news diffused fairly rapidly in an S-shaped curve, with almost all of the adult public having heard of the event within three hours (Rogers 2003). Furthermore, Rogers (2003) summarized the existing literature, arguing that situational factors such as the location of the respondent, salience, and the time of the event are predominant factors in the diffusion of the news about the event. In such a scenario, it is critical for key agencies to respond to the crisis in a relatively fast pace, staying ahead of media diffusion of information.

The crisis communication simulation would present to students the different types of crises and allow them to pick different types of timing responses to the crises (such as immediate response, response after research, prolonged response, and nonresponse). The outcomes associated with the timing responses will teach students about the best possible timing strategies in responding to crises. Not only is the timing of the crisis response important, but equally important is developing an understanding of the possible outcomes associated with the different timing strategies (Fishman 1999). In the realm of bioterror, unplanned responses that are not timed adequately run the risk of generating a plethora of undesirable outcomes ranging from public panic to mortality.

Channel Strategy

What channels would be selected for responding to the crisis? A crisis is often marked by the presence of a plethora of possible communication channels (newspaper advertisements, press releases, television news, Internet, etc.). The choice of the channel is dependent upon the nature of the crisis and the objectives of crisis communication (Dougherty 1992; Dutta-Bergman 2003a; Dutta-Bergman 2003b; Sellnow and Seeger 2001). In order to minimize chaos and streamline crisis communication, it is critical for crisis managers to optimize the choice of communication channels and harness the power of channels to the greatest extent. The importance of strategically selecting communication channels is demonstrated in the use of media during the floods of the Red River Valley. The case of the floods demonstrated that the most effective channel for reaching out to residents who needed "to receive information literally while atop the expanded riverbanks fortifying sandbag barriers against the relentless rising water" (Sellnow and Seeger 2001) was the radio. The urgency of the floods meant that residents could not rely upon newspapers and the evening news on television to get updates about the rising water situation (Sellnow and Seeger 2001). In addition, the rising water in the communities meant that in many cases the only communication technology to which the residents had immediate access was the radio (Sellnow and Seeger 2001). Radio served as a center for coordinating crisis communication, updating citizens about the latest situation, coordinating human resources, and providing timely information based on call-ins from residents (Hindman and Coyle 1999).

In the realm of bioterror, the choice of improper channels runs the risk of not reaching the most critical stakeholder groups with the message. The nonselection of a channel may leave misinformation campaigns unchallenged, leading to further chaos. For instance, in response to a bioterror event, personal Web sites start disseminating incorrect information about the crisis; the choice of the CDC not to monitor the Internet and use the channel for sending out its message may lead to the unchecked diffusion of the incorrect information, further contributing to chaos during the crisis. The anthrax attacks demonstrated this very situation, revealing that a large number of citizens went to the Internet to gather information about the attacks. In this situation, it was critical for key stakeholders to have a Web presence in order to reach out to the segment of opinion leaders who were gathering their information from the Internet.

Yet another aspect of channel strategy is the coordination of channels to send out a unified message. The channels chosen for the crisis response exercise ought to match with each other. In addition, communication managers need to ensure that the channels are optimized in the context of their capacities. The incorporation of too much information or extensive usage of a particular channel may lead to channel overload. For instance, the overloading of telephone hotlines created in the face of a crisis may actually cause more frustration than allaying public concern. Therefore, those channels need to be incorporated into the communication plan that have the capacity to handle the surge in inflow of information from the public.

Elements of Message Design
The pedagogical framework proposed here is driven by a strategic approach to message construction, articulating the specific elements of message design

that need to be highlighted in a particular crisis case (Dutta-Bergman 2003c). The message characteristics that play key roles in bioterror crises communication campaigns are (a) content, (b) appeal, and (c) source.

Content Characteristics

What are the content-based characteristics of effective communicative responses to crises scenarios? Content characteristics refer to the properties of the information; they qualify the nature of the information. A newly emerging line of communication research borrows quality criteria from the literature in library and information sciences to establish the characteristics of effective communicative content (Dutta-Bergman 2003c; Dutta-Bergman 2004b). A review of the extant literature on information quality suggested the following preliminary components: completeness, accuracy, recency, readability, relevance, and novelty.

The literature review provided a rough outline for the operational definition of the quality criteria for the purpose of constructing the messages (Dutta-Bergman 2004a). Quality is operationalized in the following manner: *completeness* (extent to which all necessary elements are included in the information), *accuracy* (degree of agreement of the information with the information provided by the best evidence or with the generally accepted scientific practice), *recency* (extent to which the information is current and covers the latest discoveries of the field), *readability* (extent to which the information can be deciphered and understood), *relevance* (extent to which the information is applicable to the user), and *novelty* (extent to which the information is new and differs from existing information). Each of these elements plays a critical role in bioterror communication because of the tremendous amount of uncertainty brought forth by such events on the one hand and the importance of public response based on clearly communicated information on the other hand. Health organizations have a critical role to play in addressing the public's need for information by providing high-quality information.

Let us take, for instance, the role of information completeness in crisis communication. In recently published research, Dutta-Bergman (2003a; 2003b; 2004a) documents the role of completeness in the creation of high quality messages. Merriam Webster's Collegiate Dictionary defines completeness as "having all necessary parts, elements, or steps." Complete information includes all the elements that are necessary to establish it. These necessary elements are agent/theory and method (Boller et al. 1990). Agent/theory targets the specific linkages that build the argument and support the claim, and method addresses how the data were gathered. Boller et al. borrowed the Toulmin (1958) model of argument structure to elucidate agent/theory. The Toulmin model suggests that a sound logical argument contains three elements: claim assertions, evidence (grounds), and authority (warrants and backing). The presence or absence of these elements makes an argument strong or weak. A claim asserts the advantages/disadvantages of a proposed action. Evidence (or grounds) are facts presented to back the claim. Warrants provide the link between the claim and the presented evidence, often serving as explanations. Backing, in turn, is the factual element that supports the warrant.

When faced with bioterror, citizens need to have information that is comprehensive in its coverage (Coombs 1995; Coombs 1999). Complete information not only provides the citizens with answers to the why and how of the crisis but also documents the steps to be taken, the preventions to be put in place, and the limitations of the current state of crisis-related knowledge (Dutta-Bergman 2003a; Dutta-Bergman 2003b; Dutta-Bergman 2004a; Dutta-Bergman 2004b). For instance, during its communication about the anthrax attacks, the Centers for Disease Control and Prevention (CDC) did not clearly communicate to the public the extent of uncertainty and the limits of its pronouncements (Babrow and Dutta-Bergman 2003). Such strategies, it may be argued, are much more likely to add to the existing uncertainty rather than mitigating it. Such assessments of information based on criteria may also be extended to other indicators of quality.

Appeal Characteristics

Selecting a clear and distinct appeal strategy lies at the core of effective communication; this is particularly important during crises. A clearly defined image enables consumers to identify the needs satisfied by an organization and serves as a key to organizational success (Roth 1995). This is especially the case during crises because the uncertainty generated by a crisis threatens the very credibility of the organization. As Roth (1995) argues, developing a needs-based image strategy provides the foundation for marketing program development and enables the organization to create a clear and distinct position within its category. Being able to clearly position itself as a critical player during times of crisis contributes to the credibility and value of the organization; the clear positioning is intrinsically tied in with the presentation of a clear image consistent with the needs of the public.

Although the literature demonstrates a large number of classifying systems for appeal types, these systems share common underlying themes. The normative model (Roth 1995) is one of the most comprehensive and most widely used models in current literature and captures the major elements in other models of appeal strategy. According to the normative model, appeal types may be classified as functional, social, and/or sensory.

A functional image strategy is utilitarian in nature and proposes to solve and prevent consumption-related problems (Shavitt et al. 1992). Utilitarian attitudes, focusing on the inherent qualities and benefits of the product, maximize the rewards and minimize the punishments obtained from objects in one's environment, guiding the behavior in a direction that obtains the benefits associated with the objects (Shavitt et al. 1992). Therefore, in the context of crisis communication, functional appeals focus on the rational elements of the crisis and the corresponding response. Functional appeals in the realm of bioterrorism may actually lay out the specific steps to be taken by citizens in order to protect themselves.

Social appeals fulfill internally generated needs for self-enhancement, role position, group membership and affiliation, or ego-identification (Roth 1995) clustered together as the social identity function. Such strategies are in accord with image-based attitudes that focus on the impressions created by using the product. In this context, attitudes function in the service of one's public image and self-expression. Attitudes help gain social acceptance by mediating relationships with other people. They also symbolize and express one's identity by promoting identification with reference groups. Use of social appeals is critical in bioterrorism because crises such as the anthrax attacks threaten the existence of the very community and unite members of the community by the commonality of the threat they face.

Sensory images build around the novelty, variety seeking, and sensory gratification needs. The importance of experiential needs in consumption has been illustrated by work on variety seeking, consumer aesthetics, and experiential consumption (Dutta-Bergman 2003b). Therefore, sensory appeals in crisis communication focus on the sensory benefits of recommended actions. Sensory appeals may focus on the emotions and feelings attached to a crisis. In the realm of bioterrorism, it is important for public health entities to appeal to human emotion, a critical element of the crisis. For instance, during the anthrax attacks of 2001, a large number of citizens reported feeling depressed during the period of the crisis. Although much of the information-based response of large-scale organizations typically ignores this affective aspect of an event, it might be critical to incorporate the element into communication such that public fears and anxieties are allayed.

Source Characteristics

Who is going to communicate the crisis response to the public? Extant research documents the importance of source credibility in communicating effective crisis information. A credible source is believable, and therefore, the public is more likely to put its trust in a credible source. The growing scholarly concern with source credibility relates to the extent to which consumers are getting their information from sources that are not qualified to provide the information (Dutta-Bergman 2004a; Eysenbach et al. 2002). In this context, it is important for key organizations to communicate their credibility clearly and to inform the public about ways of delineating credible sources from sources that do not have the credibility.

Trustworthiness and expertise of the source are the two critical criteria underlying source credibility judgments. Whereas trustworthiness emphasizes the intent of the source, expertise relates to the source's qualifications. A source that is not trustworthy and does not have the expertise is more likely to mislead the consumer during times of crises, leading to misdiagnosis and mistreatment. In response to bioterror crises, therefore, it is important for public health organizations to emphasize the credibility of sources to ensure adequate processing of the message.

Published scholarship points out that a plethora of other communicative strategies influence the credibility of the source. For instance, Dutta-Bergman (2003; 2004a) demonstrated that a source that does not communicate complete information is likely to be perceived as not credible. In this instance, the message communicated by an otherwise expert and trustworthy source negatively influenced the credibility of the source because it did not provide the audience with complete information. Given the centrality of source credibility in crisis communication, it is critical for organizations to be open and honest, sharing information that is readable, novel, accurate, relevant, and complete. It is also important for the organization to choose those appeal strategies that are likely to bolster its credibility in the face of the crisis.

In summary, the pedagogical framework proposed here demonstrated the importance of communication in the management of crises. More importantly, it pointed out the different elements of communication strategy that influence the effectiveness of organizational response to the crisis. The incorporation of these elements into the curriculum is critical for the preparation of learning materials related to bioterror crises communication training. Scientists and social scientists alike need to learn the best ways of communicating during crises scenarios such that they can plan well thought out strategies that respond to the very nature of the crisis.

In order to facilitate a better understanding of what is being proposed, the following use case scenario is presented.

▶▶ Typical Use Case Scenario

Let us assume a scenario where a student in the role of first-responder might experience a virtual environment simulating ground zero of a biological attack. The student might then have to gather evidence in a variety of manners, such as visual exploration of the surrounding area and conversing with witnesses who may (or may not) have relevant knowledge that could guide an appropriate response to the attack. The student will learn how to best communicate with these witnesses to obtain important information in a timely manner. From the design standpoint, because a large number of variables and data points are involved, the design team must first identify this scenario and then place them as objectives into a learning object.

Once the objective has been defined, a series of basic digital assets required for this scenario needs to be assembled. In this case, it may be sample pictures of the area that was affected or sound bites that may be associated with such an event. The 3D visualizations of the environment must be accurate enough to simulate the real environment as closely as possible. Once the quality of the ingredient elements are carefully planned out, the actual 3D visualizations used in the simulation are designed.

The student could then communicate with another student located elsewhere, who is playing the role of a response coordinator, such as a Centers for Disease Control (CDC) staff member. This second student would then have to interpret the relayed information, along with information received from other sources, and determine an appropriate response, such as holding a press conference. The student would also have to decide what information should be communicated (such as the type of biological contaminant believed to have been used) and how the information would be best conveyed. Based on this response, the students could experience the results of their actions. For example, if too little information was made public, the first responder could witness increasing numbers of curious onlookers attempting to enter the danger area and exposing themselves to the contaminant. Meanwhile, the CDC worker would begin receiving reports of increasing numbers of potential victims, and experience difficulty coordinating responses from on-site responders as they are forced to spend more time on crowd control and are unable to move needed resources such as medical personnel into the area. Conversely, release of too much information (or information released in a poor manner) to the general public could result in panic, where the first responder must deal with people attempting to flee the area and break quarantine, and the CDC worker must contend with the infection rapidly spreading to more distant areas.

The ability repeatedly to present various scenarios is a serious concern in such high-stakes training situations. Under the framework proposed here, once the students have completed an exercise, they could choose to repeat it making different selections and see how the outcomes might change. Such learning experiences would be difficult and extremely expensive to replicate in the real world, and the instructor would not have nearly as much control over what the outcomes would be—making it difficult to ensure that the students would learn or even observe the desired lesson.

One significant characteristic of a learning object is that it contains components that actively track student progress during the entire process. Appropriate feedback options are automatically selected by the learning object engine that is built into the simulation. The learner will also receive

Figure 2
Measured Response © simulation control showing scientist donning 3-D glasses (Source: Arangarasan and Chaturvedi, 2004)

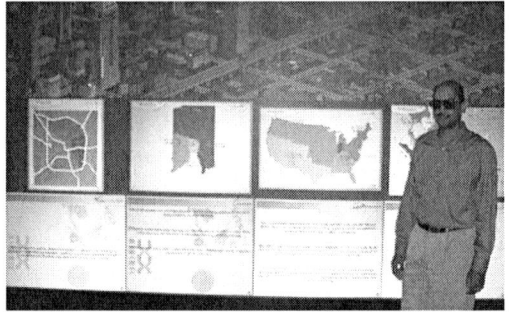

feedback from the instructor based on a full-fledged postsimulation analysis. The scores and the resulting feedback will be fed in a secure manner back to the student grade book, which is a part of the enterprise course management system. Because the entire simulation is compliant with global e-learning specifications, instructors at other universities could easily use the learning object and feed content back to the course management system at their respective universities or organizations. Figure 2, from a homeland security simulation entitled Measured Response (Arangarasan and Chaturvedi 2004) conducted at Purdue University by the Purdue Homeland Security Institute in collaboration with the Envision Center for Data Perceptualization, provides one possible conceptualization of the learning environment. However, it must be noted that some e-learning components need to be added to this environment to make it a valuable learning tool for students and in regular courses.

▶▶ Conclusion

The framework proposed here addresses fundamental issues of homeland security. In developing educational materials that address bioterror crises communication, this framework provides much needed perspectives on preparing future students to work in environments where split-second decisions can make the difference between life and death for thousands of people. The impact of this framework will in the long run translate to significantly better handling of bioterror crises communication scenarios. While the authors sincerely hope that such skills would never be called upon, as this would signal a bioterror crises scenario, given the volatile nature of the world and the epidemic spread of terrorism, it is critical to anticipate the need for crises handling skills. The educational materials developed using this framework can be used by several national agencies such as the Centers for Disease Control (CDC) and the Department of Homeland Security that may require such training. In integrating cutting-edge research with educational materials development, this chapter brings out the need for more such efforts to build crossdisciplinary relationships in dealing with complex problems. The responsibility for developing student skills consistent with national and international trends is of primary importance. Simultaneously, there is a clear need for highly trained personnel in the area of national bioterror defense and response. This framework addresses this issue in a non-trivial way. The primary responsibility for homeland defense is not the realm of the Department of Homeland Security or the Department of Defense alone. Educators, students, parents, and members of the larger general public need to be drawn into this effort by extending this framework also to accommodate the informal education of the general public.

References

Acker, S., D. Pearl, and S. Rissing. 2003. Is the Academy Ready for Learning Objects? Chatsworth: 101 Communications, 2003, http://www.syllabus.com/article.asp?id=7886 (accessed June 21, 2005).

Advanced Distributed Learning. 2003. ADL Overview. Advanced Distributed Learning, http://www.adlnet.org/index.cfm?fuseaction=abtadl (accessed June 21, 2005)

ANSER Institute for Homeland Security. 2003. Dark Winter. Arlington: Analytic Services, 2003, http://www.homelandsecurity.org/darkwinter/index.cfm (accessed December 5, 2004).

Arangarasan, R., and A. Chaturvedi. 2004. Measured Response Simulations. Homeland security simulations conducted at Purdue University, West Lafayette, IN.

Arns, L., D. Cook, and C. Cruz-Neira. 1999. The Benefits of Statistical Visualization in an Immersive Environment. *Proceedings of the IEEE Virtual Reality Conference*, 88–95, Houston, TX

Babrow, A., and M. J. Dutta-Bergman. 2003. Constructing the Uncertainties of Bioterror: A Study of U.S. News Reporting on the Anthrax Attack of Fall, 2001. In C. B. Grant, ed. *Rethinking Communicative Interaction: New Interdisciplinary Horizons*. John Benjamins Press.

Billhardt, B. 2004. The Promise of Online Simulations, http://www.clomedia.com/content/templates/clo_feature.asp?articleid=382&zoneid=29 (accessed June 21, 2005).

Boller, G. W., J. L. Swasy, and J. M. Munch. 1990. Conceptualizing Argument Quality via Argument Structure. *Advances in Consumer Research* 17:321–328.

Brown, J. S., A. Collins, and P. Duguid. 1989. Situated Cognition and the Culture of Learning. *Educational Researcher* 18(1):32–42.

Burdea, G., and P. Coiffet. 1994. *Virtual Reality Technology*. New York: John Wiley & Sons, Inc.

Chi, M. T. H., P. J. Feltovich, and R. Glaser. 1981. Categorization and Representation of Physics Problems by Experts and Novices. *Cognitive Science* 5:121–152.

Coombs, W. T. 1995. The Development of Guidelines for the Selection of the "Appropriate" Crisis Response Strategies. *Management Communication Quarterly* 4:447–476.

Coombs, W. T. 1999. *Ongoing Crisis Communication: Planning, Managing, and Responding*. Thousand Oaks, CA: Sage Publications.

Cruz-Neira, C. 1995. Projection-Based Virtual Reality: The CAVE and Its Applications to Computational Science. Ph.D. thesis, Chicago: University of Illinois.

Cruz-Neira, C., D. J. Sandin, and T. A. DeFanti. 1993. Surround-Screen Projection-Based Virtual Reality: The Design and Implementation of the CAVE. *Proceedings ACM SIGGRAPH*, 135–142.

de Jong, T., W. R. van Joolingen, J. Swaak, K. Veermans, R. Limbach, S. King, and D. Guerghian. 1998. Self-Directed Learning in Simulation-Based Discovery Environments. *Journal of Computer Assisted Learning* 14:235–246.

Dick, W. 1991. An Instructional Designer's View of Constructivism. *Educational Technology* 31(5):41–44.

Dick, W., and L. Carey. 1990. *The Systematic Design of Instruction*. 3rd ed. Glenview, IL: Scott Foresman.

Dougherty, D. 1992. *Crisis Communications*. New York: Walker and Company.

Dutta-Bergman, M. J. 2003a. The Linear Interaction Model of Personality Effects in Health Communication. *Health Communication* 15(1):101–115.

Dutta-Bergman, M. J. 2003b. Health Communication on the Web: The Roles of Web Use Motivation and Information Completeness. *Communication Monographs* 70:264–274.

Dutta-Bergman, M. J. 2003c. A Comparison of the Credibility of the Sources of Health Information. *Journal of Medical Internet Research* 5(3):e21, http://www.jmir.org/2003/3/e21 (accessed June 21, 2005).

Dutta-Bergman, M. 2004a. The Impact of Completeness and Web Use Motivation on the Credibility of e-Health Information. *Journal of Communication* 54:253–269.

Dutta-Bergman, M. 2004b. Health Attitudes, Health Cognitions, and Health Behaviors among Internet Health Information Seekers: Population-Based Survey. *Journal of Medical Internet Research* 6:e15, http://www.jmir.org/2004/2/e15/ (accessed June 21, 2005).

Eylon, B. S., M. Ronen, and U. Ganiel. 1996. Computer Simulations as Tools for Teaching and Learning: Using a Simulation Environment in Optics. *Journal of Science Education and Technology* 5(2):93–110.

Eysenbach, G., J. Powell, O. Kuss, and E. Sa. 2002. Empirical Studies Assessing the Quality of Health Information for Consumers on the World Wide Web. *Journal of the American Medical Association* 287:2691–2700.

Fishman, D. A. 1999. ValuJet Flight 592: Communication Theory Blended and Extended. *Communication Quarterly* 47:345–375.

Gagne, R. M., L. J. Briggs, and W. W. Wagner. 1992. *Principles of Instructional Design.* 4th ed. Orlando, FL: Harcourt, Brace, Jovanovich.

Glaser, R. 1992. Expert Knowledge and Processes of Thinking. In D. F. Halpern, ed. *Enhancing Thinking Skills in the Sciences and Mathematics*, 63–65. Hillsdale, NJ: Erlbaum.

Hindman, D., and K. Coyle. 1999. Audience Orientations to Local Radio Coverage of a Natural Disaster. *Journal of Radio Studies* 6:8–26.

Hodgins, H. W. 2000. The Future of Learning Objects. In David A. Wiley, ed. *The Instructional Use of Learning Objects*, http://reusability.org/read/chapters/hodgins.doc (accessed June 21, 2005).

IEEE LTSC. 2004. *WG12: Learning Objects Metadata*. Piscataway: IEEE, http://ltsc.ieee.org/wg12/index.html (accessed June 21, 2005).

IMS. 2001. *IMS Global Learning Consortium Inc.* Burlington: IMS Global Learning Consortium, 2001, http://www.imsglobal.org (accessed June 21, 2005).

Jayakumar, S., R. G. Squires, G. V. Reklaitis, P. K. Andersen, B. C. Choi, and K. R. Graziani. 1994. The Use of Computer Simulations in Engineering Capstone Courses: A Chemical Engineering Example, the Mobil Catalytic Reforming Process Simulation. *International Journal of Engineering Education* 9(3):243–250.

Mackenzie, J. G., R. M. Allen, W. B. Earl, and I. A. Gilmour. 2001. Amoco Computer Simulation in Chemical Engineering Education. *Journal of Engineering Education* 90(3):331–345.

Madhavan, K. P. C. 2004a. Where Have the Semester Papers Gone? Impact of Course Management Systems on Traditional Classroom Assessment. *Proceedings of the International Conference on Education and Information Systems: Technologies and Applications.* Orlando, FL.

Madhavan, K. P. C. 2004b. Managing Storage on Your LMS: Learning Objects. *Syllabus* July/August:31–34.

McLellan, H. 1994. Situated Learning: Continuing the Conversation. *Educational Technology* 34(10):7–8.

Noll, M. 2003. *Crisis Communication: Lessons from September 11.* New York: Rowman and Littlefield Publishers.

Oliver, R., A. Omari, and J Ring. n.d. Connecting and Engaging Learners with the WWW, http://elrond.scam.ecu.edu.au/oliver/docs/98/TLF.pdf (accessed June 21, 2005).

Reigeluth, C. M., and E. Schwartz. 1989. An Instructional Theory for the Design of Computer-Based Simulations. *Journal of Computer-Based Instruction* 16(1):1–10.

Rogers, E. 2003. Diffusion of the News of September 11 Terrorist Attacks. In M. Noll, ed. *Crisis Communications: Lessons from September 11*, 17–30). New York: Rowman and Littlefield Publishers.

Rogoff, B. 1984. Introduction: Thinking and Learning in Social Context. In B. Rogoff and J. Lave, eds. *Everyday Cognition: Its Development in Social Context*, 1–8. Cambridge, MA: Harvard University Press.

Roth, M. S. 1995. The Effects of Culture and Socioeconomics on the Performance of Global Brand Image Strategies. *Journal of Marketing Research* 32:163–175.

Sandberg, J., and B. Wielinga. 1992. Situated Cognition: A Paradigm Shift? *Journal of Artificial Intelligence in Education* 3:129–138.

Schneider, W., H. Gruber, A. Gold, and K. Opivis. 1993. Chess Expertise and Memory for Chess Positions in Children and Adults. *Journal of Experimental Child Psychology* 56:323–349.

Seeger, M. W., T. L. Sellnow, and R. R. Ulmer. 1998. Communication, Organization, and Crisis. In M. Roloff, ed. *Communication Yearbook, 21*, 230–275. Thousand Oaks, CA: Sage Publications.

Seeger, M. W., T. L. Sellnow, and R. R. Ulmer. 2001. Public Relations and Crisis Communication: Organizing and Chaos. In R. L. Heath, ed. *Public Relations Handbook*, 155–166. Thousand Oaks, CA: Sage.

Sellnow, T., and M. Seeger. 2001. Exploring the Boundaries of Crisis Communication: The Case of the 1997 Red River Valley Flood. *Communication Studies* 52:153–167.

Shavitt, S., T. M. Lowrey, and S. P. Han. 1992. Attitude Functions in Advertising: The Interactive Role of Products and Self-Monitoring. *Journal of Consumer Psychology* 1:337–364.

Spiro, R. J., P. J. Feltovich, M. J. Jacobson, and R. L. Coulson. 1991. Knowledge Representation, Content Specification, and the Development of Skill in Situation-Specific Knowledge Assembly: Some Constructivist Issues as They Relate to Cognitive Flexibility Theory and Hypertext. *Educational Technology* 31(9):22–25.

Spiro, R. J., W. P. Vispoel, J. G. Schmitz, A. Samarapungavan, and A. E. Boeger. 1987. Knowledge Acquisition for Application: Cognitive Flexibility and Transfer in Complex Content Domains. In B. K. Britton and S. M. Glynn, eds. *Executive Control Processes in Reading, 31*, 177–199. Hillsdale, NJ: Lawrence Erlbaum Associates.

Sutherland, I. E. 1965. The Ultimate Display. *Proceedings of the International Federation of Information Processing* 65(2).

Toulmin, S. 1958. *The Uses of Argument*. Cambridge: Cambridge University Press.

United States Institute of Peace. n.d. *For the Classroom: Simulations*, http://www.usip.org/class/simulations/index.html (accessed June 21, 2005).

Vieth, T. L., Kobza, J. E., and Koelling, C. P. 1998. World Wide Web-Based Simulation. *International Journal of Engineering Education* 14(5):316–321.

Wiley, D. A. 2003. Learning Objects: Difficulties and Opportunities, Utah State University, http://wiley.ed.usu.edu/docs/lo_do.pdf (accessed June 21, 2005).

Young, M. F., and McNeese, M. 1993. A Situated Cognition Approach to Problem Solving with Implications for Computer-Based Learning and Assessment. In G. Salvendy and M. J. Smith, eds. *Human-Computer Interaction: Software and Hardware Interfaces*. New York: Elsevier Science Publishers.

Chapter 4

Coordinating Effective Government Response to Bioterrorism

PAUL DRNEVICH,[1] SHAILENDRA MEHTA,[1] AND ERIC DIETZ[2]
[1]*Purdue University*
[2]*Indiana Department of Homeland Security*

▶▶ Introduction

Government organizations exist in environments characterized by varying levels of turbulence and ambiguity that have been accentuated and exacerbated by the current geopolitical climate. Although the environmental implications and nuances may differ from those experienced by private firms in conventional research settings, the assumptions upon which much of the research in the management field is grounded may be equally applicable to government organizations and their unique and extreme operating environments. A core task of organizations from a management perspective is the creation and/or maintenance of a fit between internal strengths and capabilities and the demands of their environments. Likewise, so too must government organizations draw upon unique resources and capabilities across levels to respond to issues in their environments. Similarly, the levels of turbulence and ambiguity present in a government agency's operating environment are also direct contributors to the difficulties inherent in strategic decision-making under these conditions. How these contingencies are met by government organizations and how agencies respond to the extreme environmental demands unique to managers in these environments are important issues for study. Better understanding of these issues can add much to our ability to answer some fundamental questions of management: Does strategy matter, and what are the implications of strategy choice and execution in extreme environments?

The perceived threat of terrorism in the U.S. homeland has increased dramatically this decade. Since the terrorist attack on the United States on September 11, 2001, governmental agencies at the federal, state, and local levels now face unique and extreme pressures and must make decisions with implications of extreme magnitude. To deal with these unique and challenging new situations government agencies must develop and practice coordinated response strategies for possible terrorist strikes in the United States. Because government agencies must develop coordinated response strategies for these contingencies, their strategy-making process may be similar in some respects to those of public-sector managers (i.e., multilevel responses and application of scarce resources to respond to changes in the environment). However, the nature of the environmental pressure, turbidity, and outcome implications make this a unique and interesting setting in which to study management theory.

Prior work on this topic includes extensive armed forces training and coordination development; the development of homeland defense strategy for the White House; the development, examination, and refinement of trace vaccination strategies (Rvachev and Longini 1985); and response strategies to a smallpox attack (Kaplan et al. 2002). Further uses also include numerous academic, government, and practitioner publications on epidemiological, terrorism response, and homeland security issues. However, such prior efforts have largely focused only on government policy development or epidemiological applications. Outside of the public policy or medical fields, theoretical grounding and methodological rigor has been lacking. Because management theory clearly can be extended to, and would appear to contribute to, this context, there is a well-motivated need for academic research in this area from a management perspective.

In this chapter, we examine the response decision-making process and implications of strategic choices under conditions of uncertainty and extreme risk present in a terrorist attack. The context of government agencies jointly responding to a terrorist attack allows us to study management theory in public-sector application under such unique and extreme conditions. We give particular attention to implications of characteristics of the decision makers and the implications of communication and conflict on decision effectiveness. The chapter examines propositions for outcome implications of strategic choices and coordination issues and proceeds as follows: In the first section, we provide an introduction to the setting, context, and issues; in the second section, we offer limited discussion of related theory and prior research; in the third section, we introduce and provide an overview of our solution approach, the Measured Response simulation exercise series; in the fourth section, we develop an illustrative model and propositions for the effectiveness coordinated government responses to terrorism; in the fifth section, we discuss our methods and measures for examining the illustrative model and propositions; in the sixth section, we examine the propositions through computational experimentation methods during a live training exercise utilizing computer simulation modeling; in the seventh section, we discuss the results; and in the final section, we conclude with implications for future academic research and government practitioner response behavior.

▶▶ Prior Research and Management Theory Application

General Background Issues

In our current society, we must unfortunately accept that an attack on a U.S. city involving biological, chemical, or nuclear weapons of mass destruction is a threat of serious potential consequence. Due to the extensive intelligence and information coordination challenges, government organizations cannot expect to prevent such threats in the operating environment with certainty and must, therefore, develop response strategies in the event of such a terrorist attack. The potential scale and impact of an attack requires a coordinated response by federal, state, and local governmental agencies, and a key success factor in optimizing the success and effectiveness of response measures is preparation and training. In some of these training exercises, it was found that the government agencies were poorly prepared for such events. Drills such as TOPOFF in May 2000 and 2003, which simulated the outbreak of pneumonic plague, and Dark Winter in June 2001, which simulated a smallpox outbreak, revealed the limitations of government agency preparation and leadership for effectively responding to such potential events (Inglesby et al. 2001; O'Toole and Inglesby 2002). Furthermore, the panic and confusion as evidenced during the anthrax attacks in the fall of 2001 (Chyba 2002) reveals that both the U.S. government and other world powers lack a clear and comprehensive strategy-making process to combat bioterrorism. This chapter addresses these issues and offers insights into how such events may be more effectively modeled and practiced. Background on these related issues is drawn from the management literature for the purposes of supporting our solutions approach and discussed in the remainder of this section.

Information and Communication Issues

Organizations, whether public or private, are challenged not only in capturing and leveraging knowledge and information from experiences and the environment but also in effectively utilizing knowledge and information in strategic decision-making. In turbulent environments, incomplete information or information overload (with a shortage of actionable intelligence) and a high degree of information latency prevent organizations from receiving, understanding, or acting upon useful knowledge. In observed practice, too much and often conflicting or confusing information can lead to uncertainty and delayed response or inaction. Further, once response decisions are made, information and knowledge leakage to the press, public, or other organizations can complicate and undermine the effectiveness of a response strategy. This is further exacerbated by the typical extreme and fast-paced operating environments in which government agencies try to coordinate communication and activities among levels and functions, particularly when objectives are not clearly aligned, as is often the case.

Response Decision-Making Issues

These communication and knowledge issues are central to supporting both the strategy-making process and effective strategy implementation. Strategy-making processes are the methods and practices organizations use to interpret opportunities and threats in the environment and then make response decisions regarding the effective use of organizational resources and capabilities (Shrivastava and Grant 1985). The goal of the decision process is to obtain a fit among the key variables of environment, structure, and strategy in order to achieve optimal outcomes from the strategy choice (Hart 1992). A major challenge in optimizing a response to a terrorist attack is that each group must make use of limited information when deciding upon a response strategy. Trade-offs between response effectiveness and the need for certainty may weigh heavily when responding to a biological agent. Information coordination is critical to the decision-making process to produce an effective response under this trade-off situation. This is a complex challenge because it involves horizontal communication and decision coordination and timing across different functional groups as well as vertical coordination of information flow among multiple levels of government organizations.

Agency Conflict Issues

Communication and decision coordination challenges can be prevalent in coordinated response and require balanced trade-offs for effective responses. These factors create a ripe opportunity for agency issues among the governmental departments and between the government levels. Agency issues are rooted in agency theory developed by Jensen and Meckling (1976) to explain the idea that agents are not simple utility maximizers but are boundedly rational in that they may act opportunistically to maximize self or group interests, which may undermine or suboptimize higher level objectives. An agency relationship—as it could be applied to the context of our study—is defined as a contract under which the principals (citizens, American taxpayers, etc.) elect or engage an agent (government official) to perform some service on their behalf, which involves delegating some decision-making authority to the agent. Agency theory attempts to describe this relationship. Extensions of the concept involve agency costs and relationships to separation and control issues. The concept is used in this study to refer to the potential for agents of different government agencies, departments, or levels to behave opportunistically in favor of self or group interests. This may be evidenced as self- or agency-aggrandizing (i.e., as observed in the O. J. Simpson trial or D.C. sniper media events) or simply inherent decision-making biases (i.e., elected agents weighting public mood more heavily than do appointed agents due to reelection concerns). The result of such decisions may at best complicate or suboptimize response strategies and at worst significantly delay or derail the organization's overall strategy. For these reasons, the communication challenge is not just one of facilitating vertical, horizontal, and diagonal information flow channels but also involves careful coordination of the content and timing of communications for optimal response. The following section presents a case observation of coordinated government agency response during a bioterror training exercise to illustrate the applicability of the previously discussed theory and the resultant implications of strategy process and choice.

▶▶ Case Illustration: Measured Response to Bioterrorism

The Measured Response (MR) exercise series takes place in a synthetic environment that simulates the consequences of a bioterrorist attack on a midsized U.S. city. The objective of the exercise is to develop and analyze policies and operating procedures to manage the public mood, maintain public health, mitigate the risk of contagion, maintain orderly movement of traffic and people, and combat the attack. The May 2002 (MR02) and July 2003 (MR03) exercises enabled participants to work on key response skills, which included risk management, prioritization, communications, incident management, cooperation, and management of public mood. Key training objectives of MR included practicing resource/risk management under an unconventional crisis situation; prioritization, timing, and intensity trade-offs of response decisions and actions; emergent communication strategy development and enhancement; real-time incident management and allocation of decision-making among different levels; execution and effort coordination among different agencies and actors; and management of public mood and expectations.

MR was developed on the Synthetic Environment for Analysis and Simulation (SEAS) platform developed at Purdue University (Chaturvedi and Mehta 2002). SEAS facilitates the creation of fully functioning synthetic environments that are reflective of real-life environments in many key aspects.

The synthetic environment functions by combining large numbers of artificial intelligent agents with a relatively smaller number of human agents to capture both detail-intensive and strategy-intensive interactions. These artificial agents run on a distributed tera-scale grid computing environment. To illustrate this, in MR several hundred thousand artificial agents can be employed to mimic the behavior of the simulated population in terms of mobility, the feeling of well-being as regards security (financial and physical), health, information, and civil liberties.

In regard to the human agents operating in the MR environment, MR uses three broad classifications of government responders (federal, state, and local). The federal government response level is defined as consisting of the Department of Homeland Security (HLS), the Department of Transportation (DOT), and the Department of Health and Human Services (HHS). Although there are numerous other components of the federal government that may be involved in the communications loop, including the military, the FBI, U.S. Customs, the Coast Guard, and the Centers for Disease Control, up to and possibly including the White House, it is these specific departments that are tasked with the response obligations. The state government level in MR is composed of the department of health (HHS), the department of transportation (DOT), and the state police, state bureau of investigation, and National Guard; other state field response units are encompassed in the homeland security (HLS) designation. At the local level, hospitals, emergency medical service providers, and city/county health officials are likewise encompassed in the health and human services designations (HHS), with local level homeland security (HLS) and transportation (DOT) issues managed by city/county police and fire departments.

Measured Response operates on the SEAS platform by modeling the rate of transmission as a function of population density, mobility, social structure, and lifestyle using an explicit spatio-temporal model. It uses the movement of individuals and the exposure of susceptible individuals to infected individuals to model the spread of disease. It can model the communicability of infections both from host to host (smallpox, influenza, ebola) and from the environment to the host (anthrax). In addition to standard epidemiological parameters, such as reproductive rates of infection and disease propagation rates among individuals, MR also models the hosts and pathogens via several interrelated processes. These include age-specific susceptibility, infection propagation due to the exposure of wholly susceptible populations to a newly infectious population, and population immunity necessary to prevent the epidemic (Chaturvedi and Mehta 2002; Chaturvedi et al. 2004). For the purposes of the MR02 and MR03 training scenarios ebola and smallpox outbreaks were respectively simulated on the population of a fictitious midwestern city. Participant actions and results from the MR02 training exercise were studied to develop a model and propositions for coordinated government response effectiveness. These are developed and discussed in the following section. Observations on the performance of the model and results from the MR03 training exercise are discussed later in the chapter.

▶▶ Implications for Coordinated Government Response

To visually depict the expected relationships among the factors in coordinated government response to bioterrorism, we next develop an illustrative model. The model is derived from our theoretical expectations of influences on responder behavior, as well as from our observations of actual behavior from the MR02, TOPOFF, and Dark Winter exercises. This illustrative model is depicted in Figure 1.

In regard to the relationships among the variables in the model, the responder begins by selecting a response strategy (i.e., quarantine). This strategy choice is likely moderated by the department with which the responder is affiliated, the level of government of the responder, and the type of position of the responder. The response strategy, in turn, should be directly related to effectiveness. However, based upon the characteristics of disease propagation in the simulation, the effectiveness of a strategy should also be moderated by both the timing of the response and the intensity of the resources applied to the response. While not directly depicted in the model, we further expect that there may be some indirect associations among the responder's background characteristics and the response strategy choice, timing, and intensity variables. These proposed relationships and their potential implications are developed in more detail in the remainder of this section.

Implications of Position, Level, and Agency

Numerous communication and decision-making issues were observed between the departments, positions, and levels of government officials in MR02. Further, these agency conflicts may likely be related to role dependency and public mood (elected officials favoring responses with minimal effect on public mood,

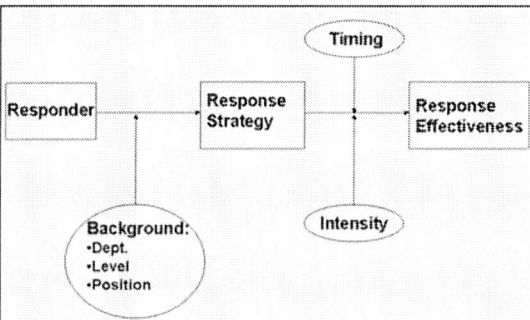

Figure 1
Coordinated Government Response Effectiveness Model

whereas non-elected agents may favor more extreme responses). Based upon the observations from the MR02 case, as well as theory predicted behavior, we make the following hypotheses regarding the implications of the position background of the government agents in the response strategy formulation and implementation process:

- *Proposition 1a*. Non-elected government officials will select more restrictive response strategies (mass quarantine/mass vaccination) than elected government officials.
- *Proposition 1b*. Non-elected government officials will have higher response intensity preferences (moderate intensity/aggressive intensity) than elected government officials.
- *Proposition 1c*. Non-elected government officials will prefer quicker responses (immediate/slight delay) than will elected government officials.

Because the decision-making process and selected response strategy may also involve the delegation of authority with regard to the proper level of government agency (federal, state, local, or a combination of actors) both in choosing and executing a response, implications for agency conflict and cross-level behavior must be considered. For example, the higher the government agency, the lower the direct accountability of the locally affected population. This should result in an increase in the restrictiveness and intensity of the strategy preference. Based upon the results of MR02, as well as theory predicted behavior, we make the following propositions regarding the implications of the level of the government agents in the response strategy formulation and implementation process:

- *Proposition 2a*. Federal government officials will select more restrictive response strategies (mass quarantine/mass vaccination) than state or local government officials.
- *Proposition 2b*. State government officials will select more restrictive response strategies (mass quarantine/mass vaccination) than local government officials.
- *Proposition 3a*. Federal government officials will have higher response intensity preferences (moderate intensity/aggressive intensity) than state or local government officials.
- *Proposition 3b*. State government officials will have higher response intensity preferences (moderate intensity/aggressive intensity) than local government officials.

Finally, among the model variables, strategy choice, intensity, and timing, government agencies involved in responding will likely differ due to their focus, priorities, and the information they utilize to formulate responses. We should, therefore, notice a difference in response strategy and intensity preference among the types of agencies responding (HLS, HHS, and DOT). Based on their focus, priorities, and information, HLS will favor more restrictive, more intense, and quicker actions than HHS and DOT. Conversely, based on different focuses, priorities, and information, HHS and DOT will likely favor less restrictive, less intense, and more delayed actions than HLS. Based upon the results of MR02, as well as theory predicted behavior, we make the following proposition regarding the implications of the types of the government agents in the response strategy formulation and implementation process:

- *Proposition 4a*. HLS officials will prefer more restrictive response strategies (mass quarantine/mass vaccination) than will HHS or DOT officials.
- *Proposition 4b*. HLS officials will prefer more intense response strategies (moderate/aggressive intensity) than will HHS or DOT officials.
- *Proposition 4c*. HLS officials will prefer quicker responses (immediate/slight delay) than will HHS or DOT officials.

▶▶ Methods

Response Variables

We collected data on the participants' preferences for certain response strategy choices in terms of quarantine, vaccination, timing, and intensity. The response variables consisted of the response strategy choice, the intensity for the strategy choice, and the timing for the strategy implementation. The set of response strategy choices can be divided into three main categories: (1) no human intervention, (2) vaccination, or (3) quarantine. These strategy choices for dealing with the bioterror attack on the city and its potential spread to other cities involve options of quarantine and vaccination response strategies. The intensity of the strategy implementation is correlated with the strategy choice. The timing of the vaccination or quarantine strategy implementation assumes no action is taken until the attack is evidenced (days 1 to 4). The response implementation can, therefore, only take place immediately after the attack (day 5), after slight delay (days 6–7), or after extreme delay (day 8). Detailed information and measures of these variables are described in Appendix A. Variables and their correlations are listed in Table 1.

Table 1. Pearson correlation coefficients

	Variable	N	Mean	S.D.	Min	Max	1	2	3	4	5	6	7
1	Quarantine	135	2.61	1.20	1.00	5.00	1.00						
2	Vaccination	135	2.52	0.98	1.00	4.00	.46**	1.00					
3	Intensity	135	2.30	0.80	1.00	3.00	.63**	.68**	1.00				
4	Timing	135	2.58	1.05	1.00	5.00	.49**	.65**	.53**	1.00			
5	Non-elected	135	0.70	0.46	0.00	1.00	0.03	0.08	0.06	0.14	1.00		
6	Federal level	135	0.33	0.47	0.00	1.00	-.15^	0.04	-.10	0.12	.29**	1.00	
7	State level	135	0.30	0.46	0.00	1.00	0.01	0.05	0.12	-.03	0.07	-.46**	1.00
8	Homeland security	135	0.33	0.47	0.00	1.00	-.02	.20*	0.03	0.03	-.06	0.00	0.06

^denotes significance at $p \leq 0.10$; *denotes significance at $p \leq 0.05$; **denotes significance at $p \leq 0.001$.

Participant Characteristic Variables

The participants in the Measured Response exercise series consisted of the government agency personnel who represented the actual responders to real events. As such, this provides a great deal of legitimacy as well as external validity to the results and the observations from this study. A variety of data was collected from the exercise participants. This consisted of the participant's department affiliation: homeland security (HLS), health and human services (HHS), or transportation (DOT). We also collected data on the government level of the participant: federal, state, or local. Additionally, we also were interested in how different types of participants might respond in these situations, so data were collected on the types of positions the participants held in their departments (elected, appointed, career employee, etc.). Beyond this descriptive data, we also collected information on the participants' preferences and priorities for decision-making. This information included assessing their preferences for trading off economic impacts for health impacts, as well as their priorities for a variety of measured factors. These factors included emergency management (ensure public safety services), energy (ensure electricity and gas services), financial (ensure financial services), food supply (ensure uncontaminated food supply), health services (facilitate efforts to promote and provide public health), health (maintain high level of public health), information (ensure availability of the information), law enforcement (ensure law and order to prevent criminal activities), mitigation (risk reduction by emergency management), public mood (maintain positive public reaction to the situation), public order (maintain high level of public order), relief (provide disaster relief assistance), transport (ensure efficient transportation flow), and water supply (ensure and regulate safety of water supply). Finally, we also collected information on the participants' perceptions of effectiveness of, satisfaction with, and confidence in their response choices, as well as the usefulness of the information they were provided to make the response decisions. Detailed information and measures of these variables are described in Appendix A. These variables and their correlations are also listed in Table 1.

Analysis Procedures

The statistical analysis consists of a multistep process. First, we conduct an exploratory analysis of the descriptive statistics of the raw, ungrouped data (results are summarized in the next section). Next, following the exploratory analysis of the raw data, we group the sample based upon combinations of the dependent variables and intensity levels for a means comparison, and analyze these to discern the significance of mean differences among the strategy, intensity, and timing alternatives. ANOVAs are then used to examine the significance of differences (results are summarized in the next section). Third, we determine the degree of the relationship among our study variables with the dependent variables in the three models (number infected and number deceased to examine our hypotheses on contagion containment and public mood score to examine our hypotheses on public mood implications). Analysis of the variance/covariance matrix from our initial analysis guided our selection of multiple regressions to attempt to create a linear combination of the independent variables to predict the dependent variables of the models. These equations are utilized with ordinary least squares (OLS) regression, beyond the ANOVA testing, as a means of exploring and examining our model and propositions. Results of this analysis are listed in the following section.

▶▶ Results

Results of the July 2003 Measured Response homeland security training exercise and the analysis of these results are presented in this section. We first present these results in terms of general observations of the outcomes of the MR03 exercise. Following this, we analyze these data to discern specific results as a means of illustrating and examining of our conceptual model and propositions.

General Observations of the MR03 Exercise Outcomes

General observations of the results of the simulation and survey data collected from the MR03 exercise indicate the following:

- Participants' weighted economic impacts were 30 percent, health impacts 70 percent, and these preferences remained fairly constant overall throughout the exercise rounds.
- Participants' initial priorities were high and remained high for emergency management, health services, health, information, law enforcement, mitigation, and public mood.
- Participants' initial priorities were moderately high or neutral but declined over the course of the exercise for energy, economy, food supply, public order, relief, transportation, and water supply.
- Participant perception of the effectiveness of the response decisions was moderately to very effective and increased throughout the exercise rounds.
- Participant perception of the satisfaction of the response decisions ranged from moderately to very satisfied and increased throughout

the exercise rounds, peaking after the third round, and declining slightly by the end of the exercise.

- Participant perception of the confidence of the response decisions ranged from moderately to very confident and increased throughout the exercise rounds.

- Participant perception of the information for the response decisions was moderate but increased throughout the exercise rounds.

- Participants were unlikely to use restrictive, mass quarantine and mass vaccination approaches, and these strategies became less likely as the rounds of the exercise progressed.

- Participants were likely to use less restrictive, city block quarantine and trace vaccination approaches initially, but these strategies became somewhat less favored, whereas city block vaccination became more likely as the exercise progressed.

- Participants chose a delayed response after some information was available, and this held constant throughout the exercise rounds, though immediate responses were favored in the early rounds of the exercise.

- Participants preferred an aggressive response, and this held constant through the exercise rounds.

Specific Results of the Analysis of the MR03 Data

Specific results from our examination of the combined survey and simulation data from the MR03 exercise are listed in Tables 2 and 3. Following this, we discuss these results in reference to our specific propositions. In reference to our specific propositions, the results from our analysis are summarized in Table 4 and discussed in further detail in the remainder of this section.

Proposition 1a stated that non-elected government officials will select more restrictive response strategies (mass quarantine/mass vaccination) than elected government officials. We *fail to observe support* for this proposition in our analysis because the variable for non-elected government officials is not significant for either quarantine or vaccination strategies with respect to the excluded category (elected officials). Further ANOVA testing reveals *no significant differences* for strategy preference between elected and non-elected participants.

Proposition 1b stated that non-elected government officials will have higher response intensity preferences (moderate intensity/aggressive intensity) than elected government officials. We also *fail to observe support* for this proposition in our analysis because the variable for non-elected government officials is not significant for intensity preference with respect to the

Table 2. Response/participant characteristics regression results

	Quarantine	Vaccination	Intensity	Timing
Non-elected	−.27	−.15	−.14	−.27*
Federal level	−.57**	.10	.16	.19
State level	−.26	.13	.13	−.01
Homeland security	−.02	.42**	.06	.08
Time dummy 1	2.22**	1.87**	1.63**	1.96**
Time dummy 2	1.80**	1.93**	1.61**	1.98**
R^2	.52	.65	.68	.59
F	22.65**	40.16**	46.06**	30.98**
Df	128	128	128	128

*denotes significance at $p \leq 0.05$;   **denotes significance at p 0.001.

Table 3. Response/participant characteristics ANOVA results

F Statistic/Significance	Quarantine	Vaccination	Intensity	Timing
Non-elected	0.13	0.82	0.45	2.69^
Federal Level	3.03^	0.24	1.48	1.93
State Level	0.01	0.39	2.11	0.14
Homeland Security	0.04	5.72*	0.14	0.12

^denotes significance at $p \leq 0.10$;   *denotes significance at $p \leq 0.05$.

excluded category (elected officials). Further, supplemental ANOVA testing reveals *no significant differences* for intensity preference between elected and non-elected participants.

Proposition 1c stated that non-elected government officials will prefer quicker responses (immediate/slight delay) than elected government officials will. We *observe support* for this proposition in our analysis because the variable for non-elected government officials is both negative and significant for timing preference with respect to the excluded category (elected officials).

Proposition 2a stated that federal government officials will select more restrictive response strategies (mass quarantine/mass vaccination) than state or local government officials. We *observe partial support* for this proposition in our analysis because the variable for non-elected government officials is highly significant for quarantine preference but not for vaccination preference with respect to the excluded category (local level officials).

Table 4. Comparison of results

Proposition	#	Response Differences	
		Mass Quarantine	*Mass Vaccination*
Non-elected officials choose more restrictive response strategies	1a	Not supported	Not supported
Non-elected officials have higher response intensity preferences	1b	Not supported	Not supported
Federal officials choose more restrictive response strategies	2a	Supported	Not supported
State officials choose more restrictive response strategies	2b	Not supported	Not supported
HLS officials choose more restrictive response strategies	4a	Not supported	Supported
		Quicker Response	*Higher Intensity*
Non-elected officials will have quicker responses than elected officials	1c	Supported	–
Federal officials have higher response intensity than state officials	3a	–	Not supported
State officials have higher response intensity than local officials	3b	–	Not supported
HLS officials have higher response intensity preferences	4b	–	Not supported
HLS officials will prefer quicker responses than HHS or DOT officials	4c	Not supported	

Proposition 2b advocated that state government officials will select more restrictive response strategies (mass quarantine/mass vaccination) than local government officials. We *fail to observe support* for this proposition in our analysis because the variable for state level government officials is not significant for either quarantine or vaccination strategies with respect to the excluded category (local level officials). Further, supplemental ANOVA testing reveals *no significant differences* for strategy preference between the levels of the officials.

Proposition 3a stated that federal government officials will have higher response intensity preferences (moderate intensity/aggressive intensity) than state or local government officials. We *fail to observe support* for this proposition in our analysis because the variable for federal level government officials is not significant for either quarantine or vaccination strategies with respect to the excluded category (local level officials). Further, supplemental ANOVA testing reveals *no significant differences* for strategy preference between the levels of the officials.

Proposition 3b advocated that state government officials will have higher response intensity preferences (moderate intensity/aggressive intensity) than local government officials. We *fail to observe support* for this proposition in our analysis because the variable for state level government officials is not significant for either quarantine or vaccination strategies with respect to the excluded category (local level officials). Further, supplemental ANOVA testing reveals *no significant differences* for strategy preference between the levels of the officials.

Proposition 4a stated that HLS officials will prefer more restrictive response strategies (mass quarantine/mass vaccination) than HHS or DOT officials. We *observe partial support* for this proposition in our analysis because the variable for HLS government officials is highly significant and positive for vaccination preference but not significant for quarantine preference with respect to the excluded category (DOT officials). Further, supplemental ANOVA testing also reveals *significant differences* for vaccination strategy preference between the HLS and HHS departments.

Proposition 4b stated that HLS officials will prefer more intense response strategies (moderate/aggressive intensity) than HHS or DOT officials. We *fail to observe support* for this proposition in our analysis because the variable for HLS government officials is not significant for intensity preference with respect to the excluded category (DOT officials). Further, supplemental ANOVA testing reveals *no significant differences* for intensity preference between the HLS and HHS departments.

Proposition 4c stated that HLS officials will prefer quicker responses (immediate/slight delay) than HHS or DOT officials. We also *fail to observe support* for this proposition in our analysis because the variable for HLS government officials is not significant for timing preference with respect to the excluded category (DOT officials). Further, supplemental ANOVA testing reveals *no significant differences* for intensity preference between the HLS and HHS departments.

▶▶ Discussion and Conclusion

Discussion of Results and Implications

We found support for propositions 1c, 2a, and 4a. This indicates the following: (1) non-elected government officials appear to prefer quicker responses than elected officials; (2) federal government level officials appear to prefer different quarantine preferences than state or local level officials; however, the effect was in the opposite direction of what we proposed in that it appears federal officials prefer less, rather than more, restrictive strategies; and (3) HLS officials appear to prefer more restrictive vaccination response strategies than HHS or DOT officials.

One of the reasons for these results may be that intense communication occurred among the participants, which quickly led to a convergence of opinion on many issues. Numerous communication and decision-making issues were observed between the departments as well as among the levels of government. Communications were fairly open and fluid within departments at the same level and within levels of government but were significantly challenged and chaotic across levels. Lengthy debates often occurred within departments and levels before action could be agreed upon and then executed. One creative result of this was that "conference call" briefing sessions emerged between rounds where agencies communicated their observations and actions and coordinated next steps for the following round. This process appeared to result in a significant improvement in the communications process and likely helped contain the spread of the disease.

We did not observe support for the other propositions. This is potentially due to measurement and sample limitations, which provide cause for us to rethink our design, measures, collection, and approach for future exercises to strengthen our analysis. Alternatively, they may have been caused by communications challenges and response delays; also, agency issues were observed in both exercises and appeared to create operational challenges and affect some outcomes in both MR02 and MR03. Specifically, government officials did not have a tendency to communicate early or often enough internally or externally (within and across both levels and agencies), as well as some participants who did not respond in as timely a manner as would have been ideal. Further, some participants were involved in extensive discussions of both the type of response action and the level of intensity of the action choice, though this was less evidenced in MR03 than in the prior year's exercise, MR02. We hope that through continued training through this and other similar exercise series, as well as improvements in the MR scenario and simulation environment, interlevel and interagency communications processes can be improved in the future.

Conclusions

This chapter examined managerial decision-making and implications of the strategy choices of actual government officials under conditions of high uncertainty and unique extreme exogenous pressures in the context of a coordinated response to a bioterrorism crisis situation. Research questions focused on the extension of management theory and simulation models to public sector application. In practical application, the chapter offers some evidence to indicate the potential usefulness of extension and applicability of management theory and agent-based simulation modeling techniques to the more extreme environments typified in homeland security tasked organizations. Finally, this research may be leveraged by future government actors to improve coordinated response strategies and increase the likelihood of an effective response to an actual event. Results also generate implications for further academic inquiry.

▶▶ Acknowledgements

The authors wish to acknowledge the National Science Foundation through which this research was funded in part by NSF DDDAS grant # CNS-0325846. We wish to thank the Purdue University e-Business Research Center (PERC) and the Purdue Homeland Security Institute (PHSI) for the use of their simulation environment and their extensive assistance with this study during the 2002, 2003, and 2004 Measured Response training exercises, as well as research assistants Tejas Bhatt and Chih-Hui Hsieh who assisted extensively with generating, coding, and compiling the data. We also wish to thank the members of the departments of homeland security, health and human services, and transportation from the federal, state, and local government who assisted with and participated in the Measured Response training exercises. We also wish to thank Jari Niemi and Gina Niemi of PHSI for their considerable assistance with the editing and preparation of this manuscript. We further thank Steve Green and the participants of his 2003 research methods seminar for input on early initial drafts of this project and for his helpful comments on our survey instrument. Finally, we wish to thank Tom Brush, Tim Folta, Roberto Mejias, Kent Miller, Mark Shanley, and Derek Ruth, who have provided informal input throughout the course of this project.

References

Chaturvedi, A., and S. Mehta. 2002. The SEAS Simulation Environment, *Technical Reports, Purdue University*, West Lafayette, IN, 519–530.

Chaturvedi, A., S. Mehta, and P. Drnevich. 2004. Computational and Live Experimentation in Bio-terrorism Response, *Dynamic Data Driven Applications Systems*, F. Darema, ed., Kluwer Publications: Boston, MA.

Chyba, C. 2002. Toward Biological Security. *Foreign Affairs* 81 (3):122–136.

Hart, S. 1992. An Integrative Framework for Strategy-Making Processes. *Academy of Management Review* 17(2):327–351.

Inglesby, T., R. Grossman, and T. O'Toole. 2001. A Plague on Your City: Observations from TOPOFF. *Clinical Infectious Diseases* 32:436–445.

Jensen, M., and W. Meckling. 1976. "Theory of the Firm: Managerial Behavior, Agency Costs, and Ownership Structure. *Journal of Financial Economics* 3:305–360.

Kaplan, E. H., D. L. Craft, and L. M. Wein. 2002. Emergency Response to a Smallpox Attack: The Case for Mass Vaccination, *Proceedings of the National Academy of Sciences*, 6(16) 10935-10940.

O'Toole, M., and T. Inglesby. 2002. Shining Light on Dark Winter. *Clinical Infectious Diseases* 34:972–83.

Rvachev, L. A., and I. M. Longini. 1985. A Mathematical Model for the Global Spread of Influenza. *Math Bioscience* 75:3–22.

Shrivastava, P., and J. Grant. 1985. Empirically Derived Models of Strategic Decision-Making Processes. *Strategic Management Journal* 6(2):97–113.

Appendix A

Measured Response 2003 Questionnaire

This questionnaire is intended to gather information on the background of the participant, measure decision-making priorities, and assess the affects of the MR03 exercise on the participant's decision making process.

1) Which level and function of government do you represent (participant #)?
   __ Federal HLS (1 2 3)   __ Federal HHS (1 2 3)   __ Federal DOT (1 2 3)
   __ State HLS (1 2 3)   __ State HHS (1 2 3)   __ State DOT (1 2 3)
   __ Local HLS (1 2 3)   __ Local HHS (1 2 3)   __ Local DOT (1 2 3)

2) What type of position do you represent?
   __ Appointed
   __ Elected
   __ Staff/Employed/Career civil service or military
   __ Other (explain) _____

3) On the following line, where would you weight your decision-making objectives between maintaining public mood/happiness and containment of the outbreak?

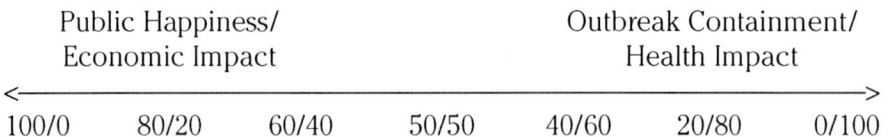

   | Public Happiness/
Economic Impact | | | | Outbreak Containment/
Health Impact |

   <——>
   100/0   80/20   60/40   50/50   40/60   20/80   0/100

4) Please rate the following priorities in terms of importance in your decision-making process:
   1 (unimportant)   2 (somewhat important)   3 (neutral)
   4 (important)   5 (extremely important)
   __ Emergency Mgmt   (Ensure public safety services)
   __ Energy   (Ensure electricity and gas services)

__ Financial   (Ensure financial services)
__ Food Supply   (Ensure uncontaminated food supply)
__ Health Services   (Facilitate efforts to promote and provide
                                  public health)
__ Health   (Maintain high level of public health)
__ Information   (Ensure availability of the information)
__ Law Enforcement   (Ensure law and order to prevent criminal
                                  activities)
__ Mitigation   (Risk reduction by emergency management)
__ Public Mood   (Maintain positive public reaction to the
                                  situation)
__ Public Order   (Maintain high level of public order)
__ Relief   (Provide disaster relief assistance)
__ Transport   (Ensure efficient transportation flow)
__ Water Supply   (Ensure and regulate safety of water supply)

*Reflecting on the round of the simulation you just completed, answer the
following questions:*

1) How EFFECTIVE do you feel your response decisions were (circle one)?

   1 (ineffective) 2 (minimally) 3 (moderately) 4 (very) 5 (extremely)

2) How SATISFIED are you with your response choices (circle one)?

   1 (unsatisfied) 2 (minimally) 3 (moderately) 4 (very) 5 (extremely)

3) How CONFIDENT are you with your response choices (circle one)?

   1 (not confident)   2 (minimally)   3 (moderately)   4 (very)
   5 (extremely)

4) How effectively were you able to integrate and process the INFORMATION
   provided in the simulation?

   1 (ineffective) 2 (minimally) 3 (moderately) 4 (very) 5 (extremely)

5) In order of importance, what three pieces of information did you rely
   on most to make decisions?

   1)

   2)

   3)

6) Do you now feel that you have successfully contained the outbreak at this time (circle one)?

   Yes   No

7) What do you feel should be the current threat level (circle one)?

   1 Green/Low   (low risk of terrorist attacks)
   2 Blue/Guarded   (general risk of terrorist attacks)
   3 Yellow/Elevated   (significant risk of terrorist attacks)
   4 Orange/High   (high risk of terrorist attacks)
   5 Red/Severe   (severe risk of terrorist attacks)

Answer the following questions based on the upcoming round of the simulation:

8) Do you feel you require more information (M) or less information from any of the following groups to improve your responses going forward (indicate by marking any applicable groups with M or L)?

   __ Federal HLS   __ Federal HHS   __ Federal DOT
   __ State HLS   __ State HHS   __ State DOT
   __ Local HLS   __ Local HHS   __ Local DOT

9) On the following line, where would you weight your decision-making objectives for the next round between maintaining public mood/happiness and containment of the outbreak?

   Public Happiness/   Outbreak Containment/
   Economic Impact   Health Impact

   <——>
   100/0   80/20   60/40   50/50   40/60   20/80   0/100

10) Please rate the following priorities in terms of current importance in your decision-making process:

   1 (unimportant)   2 (somewhat)   3 (neutral)   4 (important)
   5 (extremely important)

   __ Emergency Mgmt   (Ensure public safety services)
   __ Energy   (Ensure electricity and gas services)
   __ Financial   (Ensure financial services)
   __ Food Supply   (Ensure uncontaminated food supply)
   __ Health Services   (Facilitate efforts to promote and provide
                         public health)

__ Health   (Maintain high level of public health)
__ Information   (Ensure availability of the information)
__ Law Enforcement (Ensure law and order to prevent criminal activities)
__ Mitigation   (Risk reduction by emergency management)
__ Public Mood   (Maintain positive public reaction to the situation)
__ Public Order   (Maintain high level of public order)
__ Relief   (Provide disaster relief assistance)
__ Transport   (Ensure efficient transportation flow)
__ Water Supply   (Ensure and regulate safety of water supply)

11) Please rate your potential response preferences for the next round on the items below with the following scale:

1 (never)   2 (unlikely)   3 (neutral)   4 (likely)   5 (definite)

Quarantine preference:

__ Mass Quarantine (quarantining 100% of the population in all the geographic locations)
__ City Block Quarantine (quarantining 100% of the population in a particular geographic location)
__ No Quarantine

Vaccination preference:

__ Mass vaccination (vaccinating 100% of the population in the all the geographic locations)
__ City Block vaccination (immunizing 100% if the population of a particular geographic location)
__ Trace vaccination (trace the infection and model a vaccination strategy to "chase" the contagion)
__ No vaccination

Timing preference:

__ 1 Immediate response to limited sketchy information
__ 2 Delayed response after some information is available

__ 3 Delayed response after extensive detailed information is available

__ N No response

Intensity preference:

__ No Action

__ Moderate

__ Aggressive

Chapter 5

A Qualitative Architecture for Investigating the Quantitative Aspects of the Impact of Public Policies Related to Homeland Security

DENNIS ENGI
Sandia National Laboratories

▶▶ Introduction

Our society is faced with myriad problems that have the characteristics of being large in both scale and scope, being important to many people, possessing both social and technical aspects, requiring multiple disciplines and perspectives, and being very difficult to manage. In more technical terms, they are complex, qualitative, nonlinear, dynamic, stochastic control problems. Familiar examples include global climate change, the drug problem, the education problem, the global manufacturing network, and combating international terrorism by reducing risks to our homeland security. The quantitative aspects of homeland security risk analysis were described in *Modeling, Measuring, and Understanding Risk*, volume 1 of this series (Engi 2006). This paper describes the modeling and analysis of the more qualitative aspects of public policy impacts on the consequence dimension of homeland security.

An important element of homeland security is the protection of infrastructures that are critical to providing services to the citizenry. The electrical power and telecommunications infrastructures are well recognized as being particularly attractive targets for terrorism activities. Many of the security challenges associated with these two infrastructures are compounded by the high level of interdependency between electric power and telecommunications.

The objective of this preliminary investigation is to identify actual current (i.e., put forth in 2002 or 2003) public policy options for managing these two interdependent infrastructures to explore possible impacts on measures of quality of life. Causal relationships are investigated between public policies and quality of life related to two critical infrastructures (The White House 2002): energy and telecommunications. Specific components of these infrastructures are selected to narrow the scope of the investigation. Within the broad area of energy, the focus is the electrical power industry. The telephone and Internet communications subsystems are the focus within the telecommunications infrastructure. For the purposes of this work, these two subsystems are referred to as the telecommunications infrastructure even though a more encompassing definition would typically include broadcast television and radio, enterprise networks, amateur radio systems, and a variety of other voice and data exchange systems.

The identified causal relationships will provide a foundation for follow-on work involving the construction of a prototype dynamic system model designed to highlight both qualitative and quantitative measures of these relationships. One reason for selection of these two infrastructures is to allow the interrelationships between the electrical power and the telephone and Internet components of the telecommunications infrastructures to be modeled. A key element linking these two infrastructures is supervisory control and data acquisition (SCADA) systems, which are currently in wide use to control and monitor electrical power grids. The novelty of this research is that the interdependencies being examined are not limited to those present in the physical infrastructures introduced by SCADAs but, rather, the qualitative interdependencies that are inherent in the intertwined policy architectures. *These interdependent policy architectures give rise to complex, qualitative, nonlinear, dynamic, stochastic control problems.*

▶▶ Approach

The approach used for the investigation is based on a model architecture providing qualitative clarification of the principal cause-and-effect relationships among various policy options and categories used for determining the quality of life (Engi 1989; Engi 2000). Figure 1 shows the basic structure of the qualitative model. Six categories of policy options are included

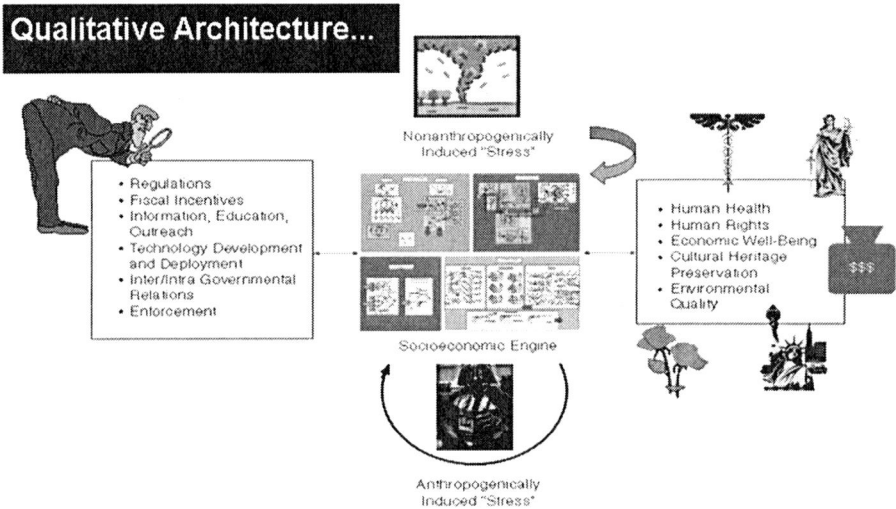

Figure 1
Qualitative Architecture for Analyzing the Impact of Policy Options on Quality of Life

(i.e., regulations; fiscal incentives; information, education, and outreach; technology development and deployment; inter/intragovernmental relations; and enforcement). The model uses five fundamental categories to characterize quality of life, namely, economic well-being, environmental quality, human health, civil liberties, and cultural heritage preservation. These *policy* and *quality of life* categories emerged from a collection of stakeholder discussions that were facilitated from 1990 through 2000 [3]. In the qualitative architecture, policy options are transformed into impacts on the quality of life through a mechanism labeled the *socioeconomic engine* (see Figure 1).

▶▶ Qualitative Architectures for the Electrical Power and Telecommunications Infrastructures

For this investigation, a small sample of public policies was abstracted from a variety of published sources fitting within the six categories of policy options included in the qualitative architecture. Sets of current policy options for the electrical power and the telecommunications infrastructures were selected to allow the modeling effort to be based on actual conditions. The two infrastructures (i.e., electrical power and telecommunications) were used to form the basis of the socioeconomic engine through which these policies impact the quality of life.

▶▶ Policy Options for the Electrical Power Infrastructure

The selected policy options for the electrical power infrastructure include:

- *Regulations*. Reduction of sulfur dioxide (SO_2), nitrous oxides (NO_x), carbon dioxide (CO_2), and mercury emissions from electric power plants (United States Senate 2003)
- *Fiscal Incentives*. Payments of up to 2.5 cents per kilowatt-hour (kwh) to owners of advanced power systems technology facilities and security and assured power facilities (United States House of Representatives 2003a)
- *Information, Education, and Outreach*. Encourage adoption of transmission technologies using real-time monitoring and analytical software (United States House of Representatives 2003b)
- *Technology Development and Deployment*. Development and demonstration of new clean coal technologies (United States House of Representatives 2003c)
- *Inter/Intragovernmental Relations*. Encourage federal agencies to participate in state or regional demand-side reduction programs (United States House of Representatives 2003d)

The term *advanced power systems technology facilities* in the fiscal incentives policy option means a facility using an advanced fuel cell, turbine, or hybrid power system or a power storage system. The term *security and assured power facilities* means an advanced power system technology determined by the secretary of energy, in consultation with the secretary of homeland security, to be in critical need of secure, reliable, rapidly available, high-quality power for critical governmental, industrial, or commercial applications.

▶▶ Causal Loop Diagrams for the Electric Power Infrastructure

Figure 2 illustrates the relationships between the five policy options for the electrical power infrastructure and the five quality of life categories. An arrow connecting a policy with a quality of life category depicts that a given policy impacts that quality of life category. As is evidenced in Figure 2, not all policy options impact all quality of life categories.

Causal loop diagrams for each of the five policy options are shown in Figures 3 through 7. Reference to Figure 3 indicates that the policy option for regulations impacts four of the quality of life categories (see also Figure 2).

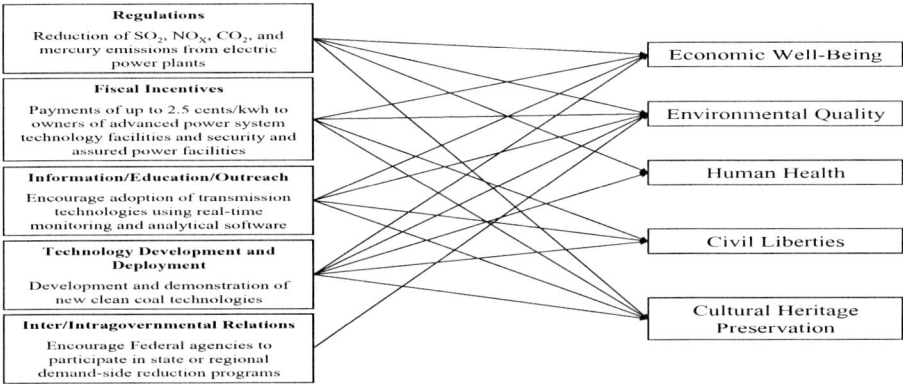

Figure 2
Relationships among the Five Policy Options and the Five Categories Measuring Quality of Life for the Electrical Power Infrastructure

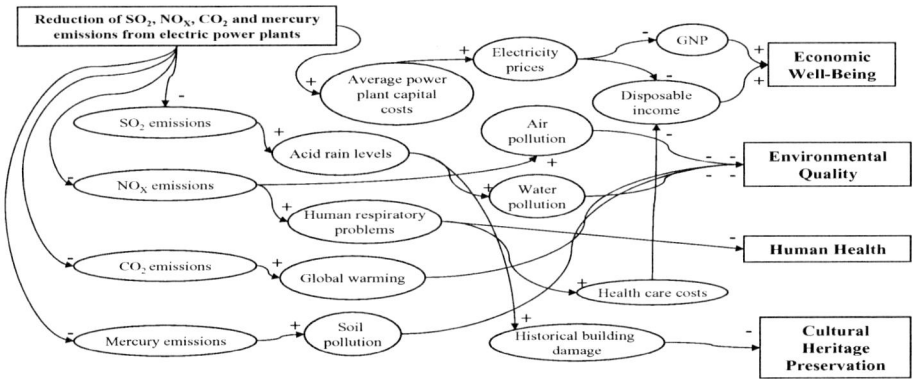

Figure 3
Causal Loop Diagram for the Regulations Policy Option for the Electrical Power Infrastructure

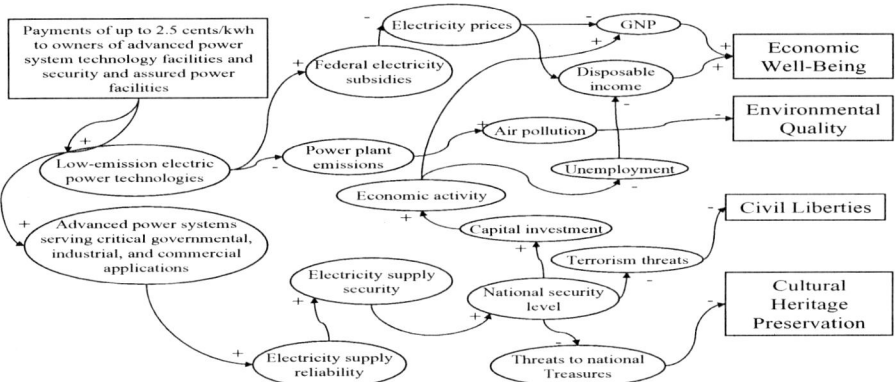

Figure 4
Causal Loop Diagram for the Fiscal Incentives Policy Option for the Electrical Power Infrastructure

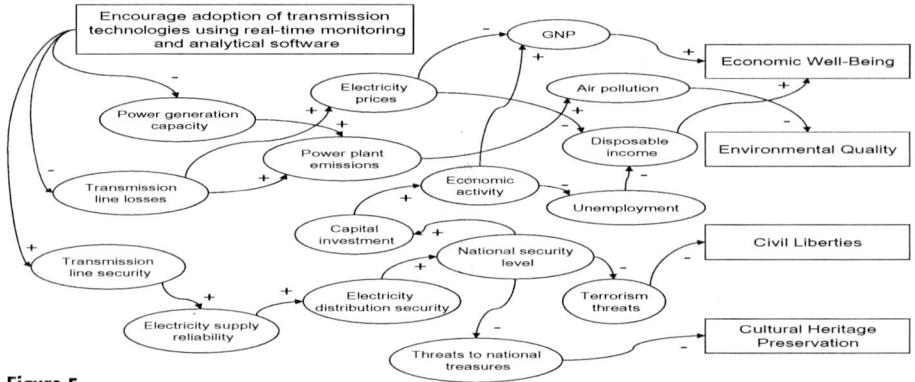

Figure 5
Causal Loop Diagram for the Information, Education, and Outreach Policy Option for the Electrical Power Infrastructure

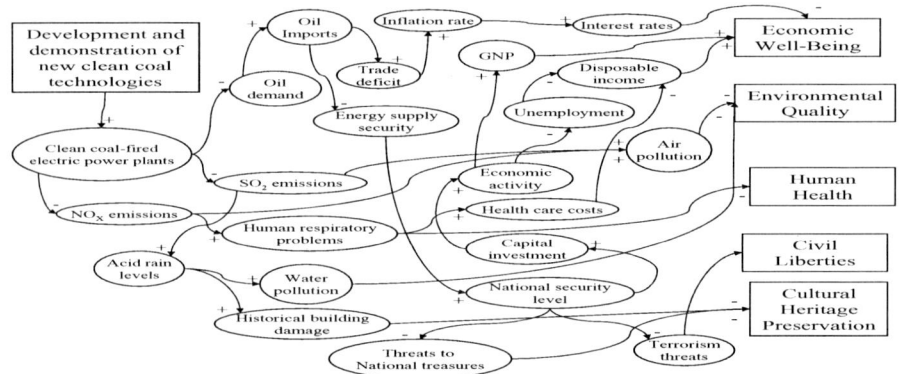

Figure 6
Causal Loop Diagram for the Technology Development and Development Policy Option for the Electrical Power Infrastructure

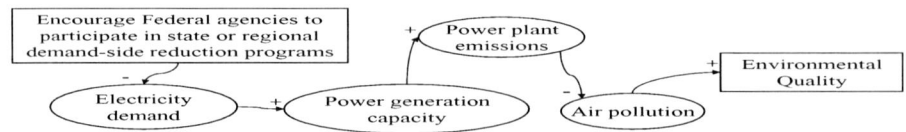

Figure 7
Causal Loop Diagram for the Inter/Intragovernmental Relations Policy Option for the Electrical Power Infrastructure

Each of the causal loop diagrams depicts a simplified illustration of the couplings between different state variables. Qualitatively, a given state variable is coupled to another state variable if an arrow connects the two variables. The sign adjacent to the head of an arrow indicates either a positive (+) or negative (−) coupling. A plus sign implies that an increase (or decrease) in one state variable causes an increase (or decrease) in the other. Conversely, a negative sign shows that an increase (or decrease) in one state variable leads to a decrease (or increase) in the other. For example, an increase in SO_2 emissions causes an increase in acid rain levels (see Figure 3). On the other hand, an increase in the deployment of clean coal-fired power plants results in a decrease in SO_2 emissions (see Figure 6).

Referring to Figure 3, implementation of the policy related to regulations decreases NO_x emissions, which in turn leads to decreased human respiratory problems, which positively impacts the quality of life expressed in terms of human health. The overall impact of a policy option on a given quality of life category expressed through a given pathway of state variables is determined by the product of the individual signs of the causal linkages. For example, requiring reduced emissions of NO_x by electric power plants increases economic well-being through the pathway of NO_x emissions to human respiratory problems to health care costs to disposable income. In this case, a minus times a plus times a second plus times a second minus times a third plus yields a plus, or a positive impact.

At the same time, this policy has a negative impact on economic well-being through the average power plant capital costs to electricity prices to gross national product (GNP) pathway because a plus times a second plus times a minus times a third plus yields a minus. However, the causal loop diagram does not suggest the net impact on economic well-being achieved through these two pathways. That result is left to the modeling effort that will have the capability to effectively assign respective magnitudes through functional relationships to the coupling between state variables, which will result in quantitative estimates of the impact on quality of life categories (e.g., economic well-being) in contrast to the purely qualitative estimates illustrated by a causal loop diagram.

The state variable greenhouse effect in Figure 3 is treated here in an exceptionally simplified manner. In actuality, the greenhouse effect is a very complex physical phenomenon that has been described in the form of a causal loop diagram previously and, therefore, is not repeated here (Engi 1989). If more detail is desired, this state variable could be replaced by such a causal loop diagram. The same observation applies to a number of state variables (e.g., power plant emissions in Figures 4, 5, and 7) identified in Figures 3 through 7. That is, expansion of a given state variable could be achieved by substituting a more detailed causal loop diagram representing the interactions of other variables associated with that state variable.

In this preliminary investigation, many of the relationships between two state variables are oversimplified. For example, in Figures 4 and 5, security levels (i.e., electricity supply security and energy distribution security) associated with the electrical power infrastructure are merely assumed to have a positive impact on the state variable national security level without providing any explanatory detail.

The causal loop diagram shown in Figure 6 is the most complex of the five causal loop diagrams illustrated in Figures 3 through 7 because more state variables appear. Some pathways involve as many as nine state variables (e.g., clean coal-fired power plants, oil demand, oil imports, energy supply security, national security level, capital investment, economic activity, unemployment, and disposable income).

The causal loop diagram shown in Figure 7 is an example of focusing the analysis on the most highly impacted quality of life category, namely environmental quality, by implementing a policy option to encourage the reduction of electricity use (i.e., energy conservation). The principal objective of demand-side programs is to reduce the need for electrical power generation capacity. However, reducing electricity demand could also result in other pathways of state variables not elaborated in Figure 7 leading to impacts on the quality of life.

▶▶ Policy Options for the Telecommunications Infrastructure

The selected policy options for the telecommunications infrastructure include:

- *Regulations.* Inclusion of qualified counterterrorism costs in the rate base of electric utilities to allow recovery of these costs (National Research Council 2002a)

- *Fiscal Incentives.* Establish incentives for investments made for security purposes in competitive market environments (National Research Council 2002a)

- *Information, Education, and Outreach.* Promote the use of best practices in information and network security throughout all relevant public agencies and private organizations (National Research Council 2002b)

- *Technology Development and Deployment.* Develop tools and design methodologies for IT systems that support graceful degradation in response to an attack (National Research Council 2002c)

- *Inter/Intragovernmental Relations.* Ensure that a mechanism exists for providing authoritative IT support to federal, state, and local agencies with immediate responsibilities for responding to terrorist events (National Research Council 2002b)

- *Enforcement.* Enhance capabilities for preventing and prosecuting cyberspace attacks (The White House 2003a)

▶▶ Causal Loop Diagrams for the Telecommunications Infrastructure

Figure 8 illustrates the relationships between the six policy options for the telecommunications infrastructure and the five quality of life categories. As in the case of Figure 2 for the electrical power infrastructure, an arrow connecting a policy with a quality of life category implies that a given policy impacts that category. As is evidenced in Figure 8, not all policy options impact all quality of life categories.

Causal loop diagrams for each of the six policy options related to the telecommunications infrastructure are shown in Figures 9 through 14. Consistent with the perspective on homeland security, all of these policy options address issues closely linked to infrastructure security and/or terrorism activities. Furthermore, the policies in three of the six categories (i.e., regulations, fiscal incentives, and technology development and deployment) have

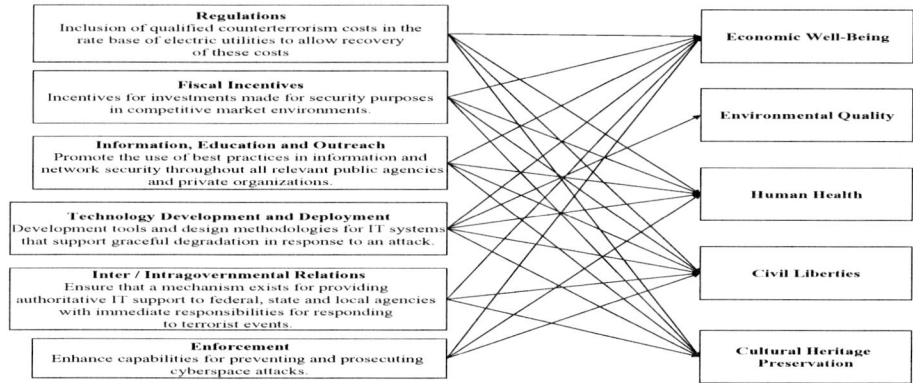

Figure 8
Relationships among the Six Policy Options and the Five Categories Measuring Quality of Life for the Telecommunications Infrastructure

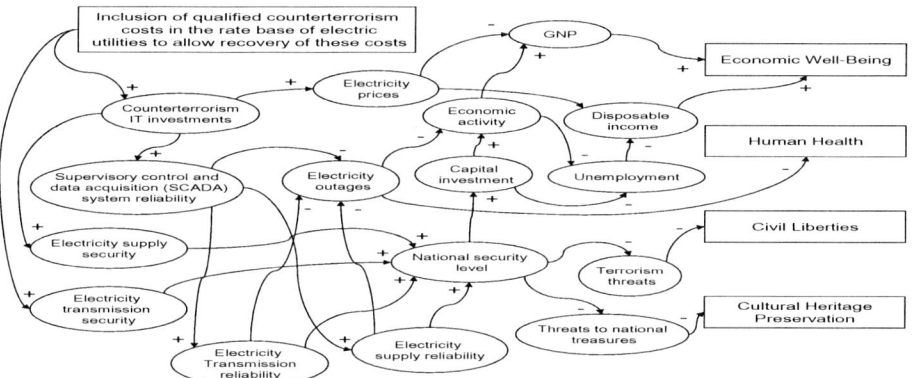

Figure 9
Causal Loop Diagram for the Regulations Policy Option for the Telecommunications Infrastructure

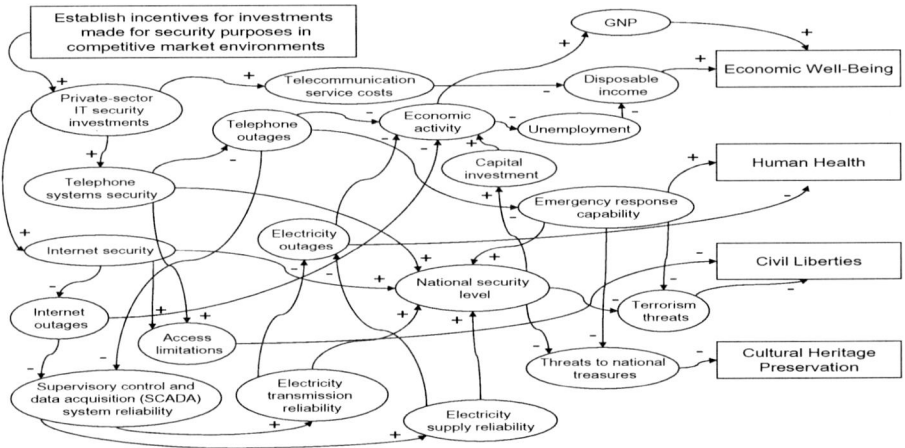

Figure 10
Causal Loop Diagram for the Fiscal Incentives Policy Option for the Telecommunications Infrastructure

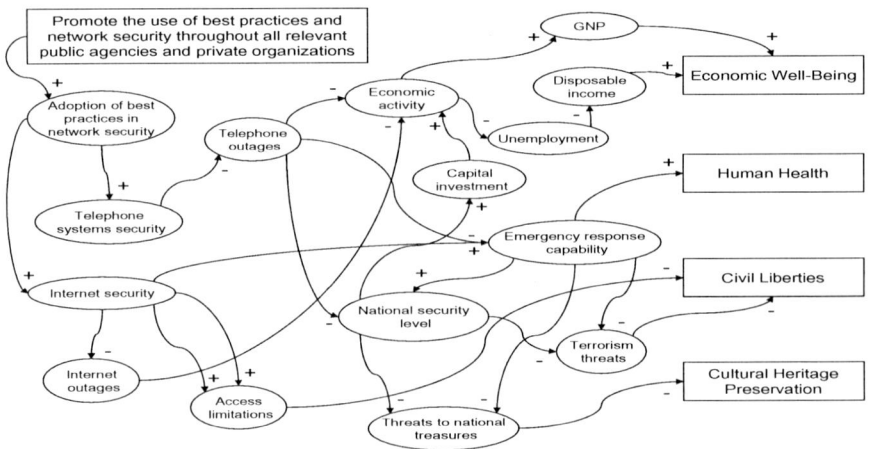

Figure 11
Causal Loop Diagram for the Information, Education, and Outreach Policy Option for the Telecommunications Infrastructure

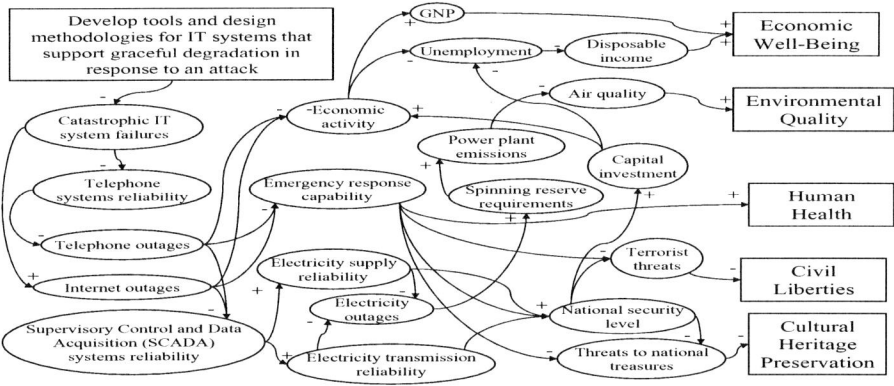

Figure 12
Causal Loop Diagram for the Technology Development and Deployment Policy Option for the Telecommunications Infrastructure

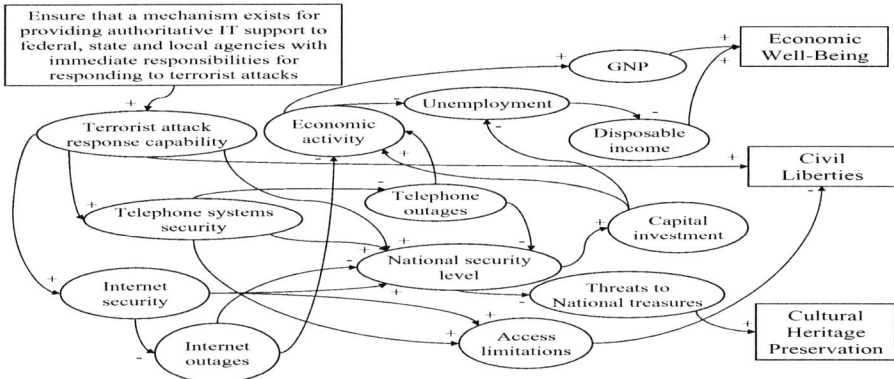

Figure 13
Causal Loop Diagram for the Inter/Intragovernmental Relations Policy Option for the Telecommunications Infrastructure

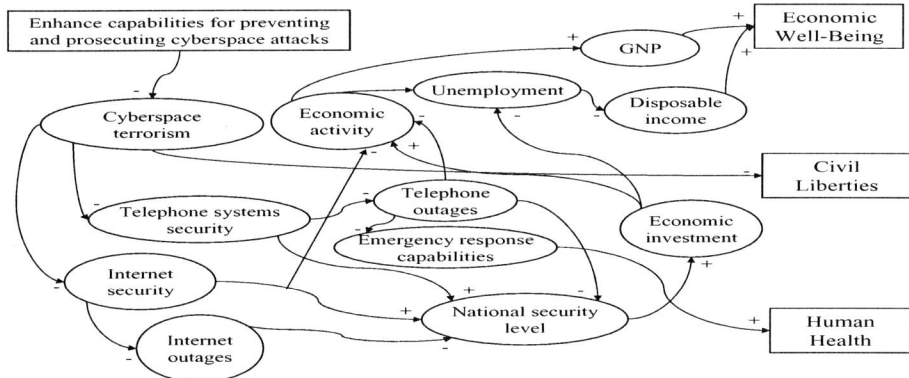

Figure 14
Causal Loop Diagram for the Enforcement Policy Option for the Telecommunications Infrastructure

direct linkages back to the electrical power infrastructure via SCADA systems (see Figures 9, 10, and 12). These choices were made intentionally to highlight the important interdependency between the electrical power and telecommunications infrastructures.

The causal loop diagrams for the telecommunications infrastructure contained in Figures 9 through 14 employ the same notation as that already explained for the electrical power infrastructure. As in the case of the electrical power infrastructure, the impact of the technology development and deployment policy option occurs through the most complex causal loop diagram.

Several of the state variables included in the telecommunications infrastructure causal loop diagrams are the same as some of those present in the causal loop diagrams for the electrical power infrastructure. This situation is particularly prevalent in the case of economic- and security-related state variables. Although shown here in separate causal loop diagrams, the modeling work will treat duplicate state variables as a single state variable. The functional duality of these variables gives rise to *policy interdependencies* that are inherent with these critical infrastructures.

Both the regulations and fiscal incentives policy options address financial matters (see Figures 9 and 10). The inclusion of counterterrorism costs in the rate structure of electric utilities would be implemented through a regulatory body such as the Federal Energy Regulatory Commission (FERC) or state-level utility commissions and, therefore, is considered as a regulations policy option. Straightforward economic incentives for competitive industries to install security equipment could be achieved by other means, for example, statutorily.

The state variable "SCADA systems reliability," which is coupled to electricity supply reliability and electricity distribution reliability, occurs in the causal loop diagrams for the regulations, fiscal incentives, and technology development and deployment policy options (see Figures 9, 10, and 12). Of course, improved SCADA reliability leads to better electricity supply and transmission reliability, both of which increase the national security level.

▶▶ Impact of an Adopted Policy Related to the Telecommunications Infrastructure

The Telecommunications Act of 1996 (United States Congress 1996) has had a profound impact on the telecommunications industry and, consequentially, on some of the categories defining the quality of life in the qualitative architecture. Of particular interest to this investigation is the detrimental impact of the act on security aspects of the telecommunications infrastructure. This situation is in marked contrast to the policies previously outlined, all of which are supportive of obtaining a more secure telecommunications infrastructure. Of course, the focus of the act was not security related. Rather, the purpose of the act was "to promote competition and reduce regulation in order to secure lower prices and higher quality services for American telecommunications consumers and encourage the rapid deployment of new telecommunications technologies" (United States Congress 1996). Nevertheless, this example illustrates how well-intentioned policy options can seek objectives that turn out to be conflicting as conditions change with time.

Among other impacts, the act effectively opened local public switched telecommunications network service to competition by requiring local exchange carriers to allow other providers of telecommunications services open access to their networks. In addition, all telecommunications carriers were required to interconnect directly or indirectly with other carriers and not to install network features, functions, or capabilities that do not comply with guidelines and standards established in the act (United States Congress 1996). In response to these mandates, carriers began to concentrate their assets in collocation facilities and other buildings, which have become known as telecom hotels, collocation sites, or peering points, rather than installing new cable. As Internet usage expanded rapidly in the late 1990s, Internet service providers (ISPs) gravitated to these facilities to reduce the costs of exchanging traffic with other ISPs (The White House 2003b).

As a result, competitive market conditions resulted in the operation of the public switched telecommunications network and the Internet becoming increasingly interconnected, software driven, and remotely managed, along with the physical assets of the telecommunications industry becoming increasingly concentrated (The White House 2003b). Unfortunately, these concentrations of key assets of the telecommunications industry could become attractive targets for terrorist activities. Furthermore, because of the growing interdependencies of certain critical infrastructures, such as telecommunications and electrical power through SCADA systems and through the required electricity to power many aspects of the telecommunications infrastructure, a direct or indirect attack on one could result in cascading effects across another. Thus, while the act was successful in increasing competition in the telecommunications industry, an unfortunate consequence is the increased vulnerability of the telecommunications infrastructure to terrorism. During the same period of time, increased competition in the electrical power industry has encouraged the reliance on public telecommunications networks and the Internet as a component of SCADA systems.

An abbreviated casual loop diagram illustrating the security-related impacts of the act is given in Figure 15. This causal loop diagram differs from those shown in Figures 3 through 7 and 9 through 14 in that the orientation is retrospective rather than prospective.

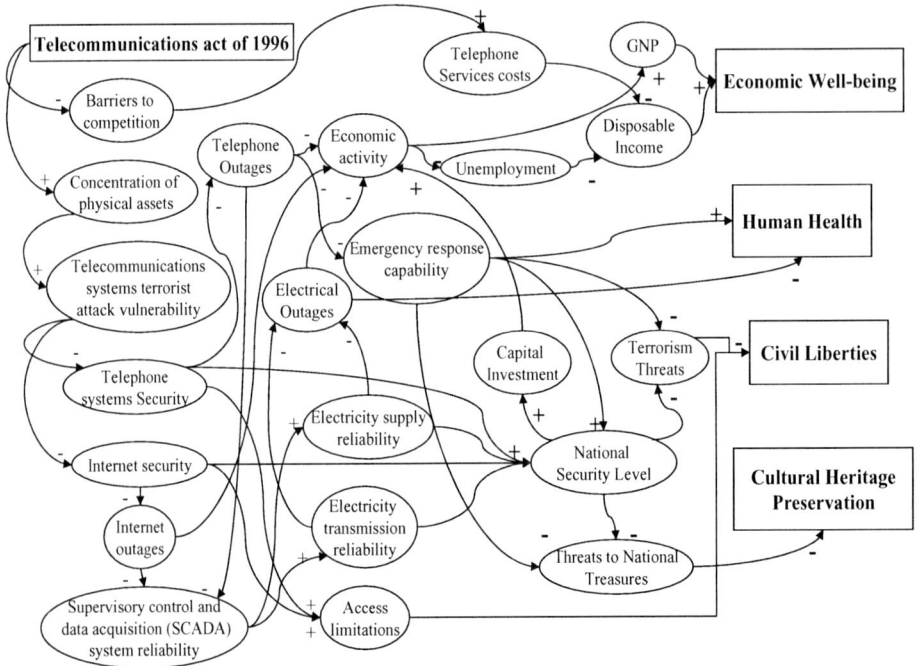

Figure 15
Causal Loop Diagram Depicting the Security-Related Impacts of the Telecommunications Act of 1996

▶▶ Concluding Remarks

The qualitative architectures detailed in this chapter and illustrated in Figures 3 through 7 and 9 through 14 for the electrical power and telecommunications infrastructures, respectively, represent highly simplified characterizations of possible real-world reactions to various policy initiatives. Nonetheless, the causal loop diagrams should be sufficiently detailed to form the foundation for the preliminary dynamic modeling effort to demonstrate the contribution that this type of analytical work can provide to understanding complex interrelationships among large numbers of variables. This type of modeling can be particularly instructive when addressing the impact of policy options on socioeconomic relationships that are commonly counterintuitive.

One way to explore the potential policy interdependencies is by examining possible combinations of policy options and quality of life categories within each of these two infrastructures. Using the six policy and the five quality of life categories as a template, there are 30 possible combinations for each infrastructure. This gives rise to 900 potential combinations or potential areas for interdependencies. Moreover, each of these 900 areas sets the context for several possible interdependent couplings. Clearly, this excessively large number of interdependencies—each of which can be inherently complex with severely stochastic, nonlinear dynamics—gives rise to a system that has yet to be understood, much less holistically managed. The next phase of this research will begin to address these compelling issues.

References

Engi, D. 1989. A Qualitative Architecture for Understanding Policy Responses to Global Change. *Proceedings of the 9th Annual Miami International Congress on Energy and the Environment.*

Engi, D. 2000. Global Approaches to Infrastructure Assurance: A Work in Progress. SAND2000-2543. Sandia National Laboratories, Albuquerque, New Mexico.

Engi, D. 2006. Modeling, Measuring, and Understanding Risk. *The Science of Homeland Security.* Vol. 1. Purdue University Press.

National Research Council. 2002a. *Making the Nation Safer: The Role of Science and Technology in Countering Terrorism.* Washington, D.C.: The National Academies Press.

National Research Council. 2002b. *Making the Nation Safer: The Role of Science and Technology in Countering Terrorism.* Washington, D.C.: The National Academies Press.

National Research Council. 2002c. *Making the Nation Safer: The Role of Science and Technology in Countering Terrorism.* Washington, D.C.: The National Academies Press.

United States Congress. 1996. Telecommunications Act of 1996, Public Law 104–104. 104th U.S. Congress, U.S. Government Printing Office.

United States House of Representatives. 2003a. Energy Policy Act of 2003, Title V, Subsection B, Section 5030A, H.R. 1664. 108th U.S. Congress, Washington, D.C.

United States House of Representatives. 2003b. Energy Policy Act of 2003, Title XII, Section 12001, H.R. 1664. 108th U.S. Congress, Washington, D.C.

United States House of Representatives. 2003c. Energy Policy Act of 2003, Title VIII, Section 8002, H.R. 1664. 108th U.S. Congress, Washington, D.C.

United States House of Representatives. 2003d. Energy Policy Act of 2003, Title I, Section 1008, H.R. 1664, 108th U.S. Congress, Washington, D.C.

United States Senate. 2003. Clean Power Act of 2003. Section 702, United States Senate, S. 366. 108th U.S. Congress, Washington, D.C.

The White House. 2002. *The National Strategy for Homeland Security*. U.S. Government Printing Office, Washington, D.C.

The White House. 2003a. *The National Strategy to Secure Cyberspace*. U.S. Government Printing Office, Washington, D.C.

The White House. 2003b. *The National Strategy for the Physical Protection of Critical Infrastructures and Key Assets*. U.S. Government Printing Office, Washington, D.C.

Chapter 6

Security Screening

J. L. FOBES, PH.D.
Transportation Security Administration
U.S. Department of Homeland Security

▶▶ Screening Checkpoint History

The requirement to submit to security screening is now commonplace in airports as well as in some government buildings, schools, and private venues. This was not always the case. The need for security screening in general very closely approximates the history of aviation security screening. Of the transportation modes under the province of the Transportation Security Administration (TSA), the overwhelming emphasis on aviation security is in part indicated by funding levels for research and development (R&D). The TSA funded aviation security R&D with 81 percent of its fiscal year 2003 R&D budget and with 80 percent of its fiscal year 2004 R&D budget (Government Accountability Office [GAO] 2004). The evolving threat described below, from Federal Aviation Administration (FAA) records (FAA 2002), is the underlying basis for the evolution of procedures and equipment at airport security checkpoints. Aviation screening was driven from its onset to respond to weapons enabling aircraft hijackings and was then expanded in scope in response to aircraft sabotage with explosives.

The airport checkpoint marked its 35th birthday in July 2005. In July 1970, the New Orleans International Airport became the first domestic American airport to require all passengers to submit to an anti-hijacking screening system. This FAA-developed process consisted of a twofold examination. People were evaluated with a behavioral profile and examined for weapons with a metal detector. This requirement was in response to numerous domestic hijackings beginning in 1961, typically undertaken to obtain transportation to Cuba. In May 1961, a passenger forced a pilot to divert to Cuba, and four more incidents occurred by the end of that August. This was to become a familiar refrain, and a rash of domestic hijackings followed. In 1968, 12 commercial and 6 general aviation aircraft were diverted to Cuba, despite the presence by then of sky marshals on some flights. The next year, 8 airliners were hijacked to Cuba in January alone, followed by an additional 23 domestic aircraft forced to land in Cuba in the remainder of 1969. The international scene was equally grim in 1969—87 hijackings occurred worldwide, 40 of which involved U.S. aircraft.

The hijacking problem was intensified in March 1970 when the first associated death occurred. In the process of a domestic hijacking, the copilot on an Eastern Airlines shuttle was shot. The first passenger death occurred the following year in a domestic hijacking of a TWA flight in June 1971. Four fugitives then killed a ticket agent before hijacking a plane to Cuba in 1972. The following month, three fugitives hijacked a plane for a 29-hour odyssey to Cuba during which the copilot was injured.

In addition to these hijackings for transportation, extortion-based hijackings began with "Dan Cooper" in 1971. This was quickly followed by more such demands, with 17 extortion attempts being made in the following six months. The upshot of these diverse hijacking episodes was that domestic carriers were required as of January 1973 to inspect all carry-on baggage for threat objects. Checked baggage would not be routinely subjected to scrutiny for years to come. Airlines were additionally required to scan each passenger with a metal detector or by a physical search. No successful hijackings occurred in the United States until 1976. However, in 1974, there was a foreshadowing of a third and much more problematic hijacking motive—using an aircraft to attack a building and kill people. A former mental patient unsuccessfully attempted to take over a DC-9 to fly it into the White House from Baltimore-Washington International Airport (BWI). Several years prior to this attempt, a presidential cabinet committee to combat terrorism identified just such a threat of someone attacking with a commercial jet used as missile. The committee raised concerns that airlines would not pay for necessary security improvements to prevent this, such as tighter screening procedures. In 1972, a White House memo stated that "The trouble with the plans is that airlines and airports will have to absorb the costs and so they will scream bloody murder should this be required of them" (Webshots 2005).

With the screening checkpoints in place for five years, 25 more attempts were made to hijack a domestic airliner. Perpetrators falsely claimed to possess weapons or explosives, but none of these actually involved successfully smuggled weapons through screening. Only one smuggling attempt was successful, when a pistol was taken through the screening process, resulting in a hijacking to Cuba in 1980. This was followed by refugees from the Mariel boatlift resorting to hijacking to return to Cuba. In 1980, 11 hijackings occurred, followed by 1 in 1981 and 3 in 1982. All but one of these involved the threat to ignite real or alleged flammable liquids. These were followed by 11 hijackings in 1983 and 3 in 1984.

Surprisingly, not many deaths resulted during hijacked flights involving U.S. carriers on domestic departures. Fatalities were limited to two flight crew members in 1970 and 1972 and a passenger in 1971. This, perhaps, lulled the government and the country at large into a false sense of security about cooperating with hijackers. This complacency ended in 2001 with the four planes comprising what has become known as 9/11. Screening for weapons to prevent hijackings was quickly broadened in scope after sabotage by explosives of the Pan Am Flight 103 over Lockerbie, Scotland, in 1988. All 259 aboard were killed, and this became a watershed event in aviation security screening. Even though the bomb was not placed aboard domestically, attention became focused on the need to detect explosives at U.S. airports. However, the threat of such sabotage did not begin with the Pan Am 103 destruction. Prior domestic examples include a bomb on a United Airlines plane that killed all 44 aboard in 1955; a Western Airlines flight surviving a dynamite explosion with the loss of only the bomber in 1957; a National Airlines plane lost with all 34 aboard in 1960; and a Continental Airlines plane blown up with a loss of all 45 aboard in 1962. Deaths from explosions averted in 1972 include a bomb discovered and defused at New York, a bomb-damaged aircraft parked at Las Vegas, and a bomb defused on a jet at Seattle.

▶▶ Equipment

Passenger Checkpoints

The requirement to intercept weapons, typically handguns used in hijackings, led to the formation of the passenger screening checkpoint. Unfortunately, security screening has typically not been approached from a systems standpoint—operators interacting with equipment to detect threats. The emphasis instead is usually on equipment development at the expense of human factors. In response to hijackings, detection equipment accordingly was chosen for metal detection during passenger inspections and x-ray machine imaging for detecting weapons in carry-on bags. The National Institute of Justice (NIJ) publishes performance requirements, testing methods, and operational guides for walk-through and handheld metal detectors (Paulter 2001a, 2002, 2003a, 2003b). Both detection portals and hand wands can be effective for assisting screeners to detect weapons such as metallic handguns and metal knives. Both sensors have evolved somewhat over time, with claims being made that portals evidence improvements in sensitivity and target location feedback. Not all of these attempts at semiautomation have been successful, however. For example, Quantum Magnetics advertises that its i-portal 100 "accurately pinpoints concealed weapons by displaying a digital picture of the person on a computer screen; threats are identified by a flashing dot on the person's image" such that "security personnel only target those highlighted areas on the body . . ." (Quantum Magnetics 2002). This did not bear up under scrutiny in laboratory testing of the portal with a variety of weapons. The i-portal 100 had difficulty correctly displaying the location of the detected item (Klock 2002).

The alarm raised by a metal detection portal merely indicates the presence of some metallic object. Resolving such an alarm requires locating the object(s) and determining whether or not they constitute a threat. This usually requires a handheld metal detector to more specifically locate the object(s) detected by the portal. For example, Fisher Research Laboratory's CW-10 claims to be sensitive enough to locate metallic objects as small as the metal springs and firing pins of composite handguns. Various performance factors have been examined for a variety of hand wands and a higher probability of detection (P_d) was associated with certain wands. However, hand wands with a high P_d were not always highly rated by the screeners for usability features (Fobes et al. 1995a). Locating the metallic item is then followed by the requirement to determine whether the item constitutes a threat. This is not always easy. For example, *The Manual of Prohibited and Concealable Weapons* (Collins 2002) lists a variety of unobtrusive weapons such as knife blades retracted within coins and belt buckles that assemble into knives.

Metal detection does not protect against explosives and nonmetallic weapons hidden on a person. Such detection requires other technologies (or strip searches), and many of these have been reviewed by the NIJ (Paulter 2001b). Among these technologies for imaging concealed threat articles are more exacting inspections of persons by equipment provided by RAPISCAN's Secure 1000 and AS&E's Bodysearch. These devices scan with an x-ray beam, some of which returns as backscatter. These returning x-rays stimulate detectors and, when scanning people, image processing algorithms then generate an image of the unclothed body. Such devices have been used in airport pilot projects (e.g., Orlando, FL [MCO]) but remain uncommon in security screening of individuals. These devices are somewhat controversial in that passengers are exposed to low-level doses of radiation and privacy concerns have been expressed. Although manufacturers can provide a cloaking device, or electronic fig leaf, not imaging some part of the body provides a place to hide contraband. Using same-sex operators might address the privacy issues. The issue is whether potential benefits offset liabilities in terms of preventing weapons smuggling for hijackings or explosives for sabotage. That is, does such imaging afford more protection than that from metal detectors? Such imaging may not do so for metallic weapons but does for nonmetallic ones, such as ceramic knives and guns, as well as explosives.

The desirability of the imaging technology becomes clearer in light of its potential to thwart attempts to smuggle explosives. In August 2004, two Russian airliners were blown up in the air after departing from Moscow's Domodedovo Airport. Passengers were believed to have smuggled aboard explosives concealed on their persons. This emphasizes the difficulty of identifying passengers carrying concealed explosives for suicide attacks and the cost of failing to do so. Backscatter technology can assist in detecting hidden explosives and is being examined for security screening of vehicles entering vehicle ferries under TSA's Secure Automobile Inspection Lanes test project. The Coast Guard is screening the vehicle (without passengers) for conventional and plastic explosives under this pilot program (Homeland Defense Journal Online 2004a). More extensive imaging of people can be obtained with devices such as the Body Orifice Security Scanner (BOSS) by Gaffco. Sometimes used with visitors to prisons, the BOSS provides noninvasive detection of metal objects hidden in oral, vaginal, anal, or nasal cavities.

The equipment of choice for inspecting carry-on bags has been the x-ray machine since the origin of the checkpoint. This machine is particularly appropriate to support detection of weapons such as assembled metallic handguns. But, as is the case with metal detectors, the operator must interpret the machine's capabilities. X-ray equipment alone is not capable of automatically detecting and identifying threat articles to a standard acceptable in the United States. The most common x-ray machine passes energy through the baggage to produce an image for the screener to interpret. After the Pan Am 103 incident, the threat from explosives became a greater checkpoint focus. Unfortunately, the x-ray image is not a particularly appropriate tool to detect what is known as an improvised explosive device (IED). In contrast to a bomb such as a hand grenade, an IED consists of the various bomb parts that can resemble everyday items. Explosives can be shaped to simulate items such as a book, and the timer can be an everyday watch along with batteries and a detonator—all of which can additionally be hidden inside something such as a piece of electronic equipment. Added to the potential to disguise an IED is the 'visual clutter' accompanying it in the typical carry-on bag and the varying densities of bags' contents. These circumstances, along with the emphasis on high passenger throughput (especially prior to the government's taking over screening), requires screeners to make judgments that are probably more difficult than those required by radiologists. While aviation screeners are not required to have a high school equivalency diploma, medical interpretation of x-ray imaging is done by highly trained individuals with no time constraints. To some extent, if you have seen one x-ray of an arm you have seen them all. Each carry-on bag has the potential to be unique.

The original transmission x-ray machines produced a black-and-white image, and a variety of attempts have been made to automate the weapons detection. One such approach for x-ray screening attempted to develop a pattern recognition program for guns and knives. While some automated success was obtained with otherwise empty carry-on bags, a representative cluttered bag precluded such automated detection for either of these weapons. Driven by the need to find IEDs, potential enhancements in x-ray display design have occurred, including color displays. Such displays typically depict organic materials (e.g., explosives) as orange and metals as blue. However, color was not found to enhance the P_d for finding IEDs under operational conditions (Fobes and Lofaro 1994; Fobes and Neiderman 2000). Realistic-appearing (for the time) IEDs were constructed with the FAA Modular Bomb Set (MBS) kits. Modular Bomb Set kits now have the capability for over 250 different combinations. These were being used to check for explosives detection capabilities at the checkpoint and contained a variety of explosive simulants along with various types of timers, batteries, wires, and a metallic detonator. Three different configurations were assembled from the kit, and four of each of these configurations were randomly presented to screeners in 12 test bags among 120 normal passenger bags at San Francisco International Airport (SFO). Screeners were asked to indicate whether they thought there might be a bomb (real or simulated) in each bag they x-rayed. This was probably one of few, if any other, such studies that also measured color vision and visual acuity. These measures did not correlate with outcomes.

Another upgrade of x-ray displays was a three-dimensional (3D) image based on binocular disparity (Image Scan Holdings, PLC's Axis-3D). Sixteen certified screeners were presented with 17 carry-on bags containing threat articles within a stream of 120 test bags. Threats included eight IEDs, four guns, three knives, and two sharp tools that would not have been allowed past the checkpoint. A comparison of detection performance with the 3D technology and conventional x-ray projection displays indicated that the 3D condition did not result in enhanced P_d. Interestingly, the 3D condition did result in a small (3%) but significant decrease in the probability of a false alarm (P_{fa}) with innocent bags (Barrientos et al. 2000a). A false alarm is said to occur if a screener reports a threat article when, in fact, there is none. Other display features offered to enhance detecting explosives are organic stripping and inorganic stripping options. Organic stripping deletes the display of organic objects while inorganic stripping, conversely, deletes inorganic objects. Neither has been demonstrated to enhance the P_d under airport operational conditions using real explosives.

X-ray equipment manufacturers offered various screener assist technologies (SAT) to somewhat automate IED detection. Four equipment versions were tested under laboratory conditions, with 12 certified screeners presented with 120 representative carry-on bags. Eight test bags contained an IED; four contained an IED and a weapon; and four contained a weapon but no IED. Screeners were tested under conditions of SAT off and SAT on using one of four machines—EG&G Astrophysics' Operator Assist System, Hiemann System's X-ray Advance Contents Tracking, Rapiscan's Auto-Detect X-ray System, and Vivid Technologies' Advanced Passenger Screening System Model APS. These machines employ dual-energy x-ray inspection in an attempt to identify materials with the mass, density, and/or atomic number signatures of explosives. The displayed image had a region of suspected explosive outlined, but the P_d for IEDs was not enhanced by these SAT (Fobes and Barrientos 1997). An even more telling failure to find a benefit with SAT was a test at Los Angeles International Airport (LAX) wherein 69 certified screeners used an EG&G Astrophysics' x-ray with operator assist turned both on and off. The SAT was examined with real explosives (as opposed to simulants of debatable representativeness) under operational conditions *in an airport.* This was done with x-ray images of IEDs made with real explosives electronically combined with the x-ray images of some carry-on baggage (Threat Image Projection [TIP], to be discussed later). The P_d was not enhanced when the SAT was active (Barrientos and Neiderman 2000).

The NIJ has also published a guide to selecting explosives detection equipment (Rhykerd 1999), including enhancing checkpoint screening for explosives with an explosive trace detector (ETD) such as GE's Ion Track Itemiser. Approximately 1,219 ETD units are being used at U.S. airports (Pryor 2004). They analyze samples taken from carry-on bags, such as laptops, for the presence of explosives. These machines are being augmented by trace portals, and the TSA is testing Smiths Detection's Ionscan Sentinel explosives detection trace portals at the Kennedy International Airport (JFK) (Homeland Defense Journal Online 2004b). Passengers walk through the portal, similar in appearance to a metal detection portal, and gentle puffs of air dislodge particles of explosives on them to obtain samples for analysis. This explosives detection portal was previously deployed in five other airports, and TSA also tested this technology with rail passengers under their Transit Rail Inspection Pilot (TRIP) project (U.S. Department of Homeland Security 2004). Passengers were screened for explosives prior to traveling on Amtrak and Maryland rail commuter trains. TSA is also testing trace machines (Smiths Detection's

Ionscan Document Scanner) that evaluate airline passenger's documents for traces of explosive residue at Reagan Washington National Airport (DCA), as well as at JFK, LAX, and Chicago (ORD) (Smiths Detection 2004). The sample, obtained by swiping the surface of documents, is evaluated in a 6- to 8-second analysis cycle. This technology was previously tested with rail passengers under TSA's TRIP project.

The final type of technology deployed for carry-on bags screening is machines using quadrupole resonance (QR), such as Quantum Magnetic's QScan QR160, to detect explosives. The technology uses pulses of low-intensity radio waves to probe molecular structures. Detection occurs when the waves result in a signal, characteristic of explosives, that is picked up by a receiver. A QScan machine was tested with real explosives and four certified screeners attempting to detect IEDs in 12 of 100 bags presented. The system

Checkpoint Screening

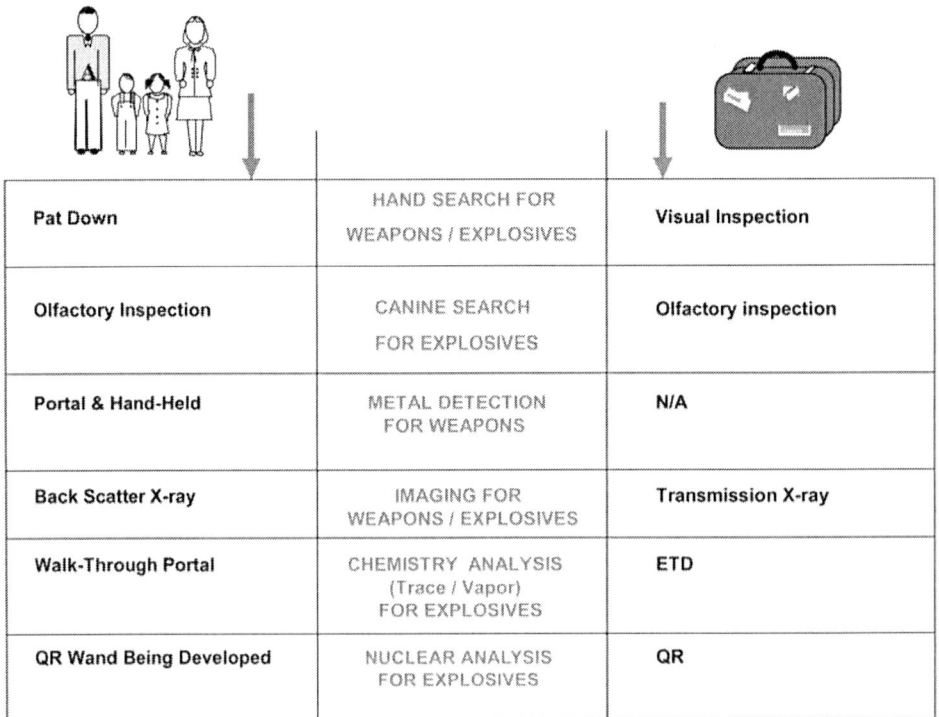

Pat Down	HAND SEARCH FOR WEAPONS / EXPLOSIVES	Visual Inspection
Olfactory Inspection	CANINE SEARCH FOR EXPLOSIVES	Olfactory inspection
Portal & Hand-Held	METAL DETECTION FOR WEAPONS	N/A
Back Scatter X-ray	IMAGING FOR WEAPONS / EXPLOSIVES	Transmission X-ray
Walk-Through Portal	CHEMISTRY ANALYSIS (Trace / Vapor) FOR EXPLOSIVES	ETD
QR Wand Being Developed	NUCLEAR ANALYSIS FOR EXPLOSIVES	QR

Figure 1

Various approaches being used to examine people and their possessions passing through the checkpoint

resulted in an increased system P_d for detecting explosives (Fobes and Monichetti 1999). The TSA is additionally in the R&D process for optimal detection techniques for screening shoes, casts and prostheses, and bottles (Polski 2004). The emphasis here is on detecting explosives. In addition to explosives, bottle screening devices (BSD) are required to detect and distinguish flammable liquids from benign ones in bottle types commonly found in the marketplace (Thompson 2004).

Checked Bags

Few checked bags were inspected prior to airline departure in the United States until after 9/11. Checked bags had been examined by x-ray for selected international destinations, but it was not until 2002 that all checked bags for domestic departures would be examined. In contrast to potential hijackings enabled by contents of carry-on bags, the threat in checked bags is not weapons but sabotage through explosives. An x-ray image of a checked bag results in an even more difficult problem for finding IEDs than with a carry-on bag image. Ten IED configurations were assembled from the MBS kit, and two of each configuration were randomly presented in 20 test bags among 200 checked bags at SFO. Only half as many IEDs were found in checked bags compared with carry-on bags (Fobes and Lofaro 1994). This is probably due to the increased size, higher clutter, and higher density of checked bags. Higher detection rates of IEDs, particularly artfully concealed ones, resulted with Isorad's SDS-400S Fluoroscopic x-ray machine, but the machine has severe usability issues (Klock, Maguire, and Snyder 2004). This technology uses an x-ray emitter two to three times more powerful than existing dual energy technology and results in a high-resolution, rotatable, virtual 3D image.

The FAA examined the potential of automated detection of explosives with testing of SAIC's Thermal Neutron Activation (TNA) machine in the late 1980s. Nuclear technology is unfortunately slow and expensive and requires more space than typically available at airports. By 1993, the FAA developed a certification standard for explosives detection systems (EDS). The standard required the *equipment* to be automated, process a minimum number of bags per hour, detect specified amounts and types of explosives, and have a minimum P_d and a maximum P_{fa}. Calling the machine a system is really a misnomer because the machine does not become a system until an operator interacts with the machine. The overall system then no longer necessarily conforms to some of the numbers in the certification standard, except for when the machine fails to detect an explosive. The computed tomography (CT) machine is typically operated by having the machine retain the bag inside and display an image of bag contents only if the CT judges explosives to be present. Thus, four machine decision outcomes are possible, and the only time the CT machine alone makes the final decision, and the operator

	Explosive Present		Explosive Absent
CT Alarm	CT Hit	Screener Yes = **SYSTEM HIT**	Screener Yes = System False Alarm
		Screener No = System Miss	CT False Alarm
			Screener No = **SYSTEM CORRECT REJECTION**
No CT Alarm	CT Miss		CT Correct Rejection

Figure 2

Possible outcomes with the presence and absence of an explosive as a function of whether or not one is judged to be present by the CT machine and/or operator

does nothing, is after a CT miss (failed to detect an explosive) or a CT correct rejection (correctly identified no explosives) as depicted in Figure 2. A CT hit (correctly identifies explosives) or a CT false alarm (incorrectly reports explosives) results in an alarm to be resolved by the operator. After a machine hit (detection), the screener can either also detect, or hit, resolving an alarm by correctly confirming the machine's detection, or miss by incorrectly over-riding the machine's detection. Concerning machine false alarms, the screener can correctly resolve the alarm by correctly rejecting the machine's evaluation or can also false alarm by incorrectly confirming a machine error. The R&D goal is to optimize *system* hits and correct rejections depicted in Figure 2 in bold italics.

Resolving the equipment alarms for potential detection of explosives is not addressed by the standard. Thus, the certification test would have been possible to pass with a green light for a bag without explosives and a red one for a bag with explosives. InVision was the first company to produce a machine that passed certification testing. To its credit, InVision provided much more than the minimum to pass by, including feedback provisions for an operator to resolve an alarm. Only three companies currently produce automated detectors (e.g., CT) that have been certified under this standard. InVision's CTX-5000 became the first EDS to be certified in 1994, later to be joined by its 5500, 2500, and 9000 machines. Discouragingly, InVision's most recently certified machine (CTX 9000DSi) has an interface that fails most of nine criteria for usability (Dixon 2003). The other companies are L3 Communications' Examiner 6000 and 3DX 1000, as well as the CT-80 machine made by Reveal

Imaging Technologies. Machines examine bag contents using density information in combination with volume information obtained by rotating an x-ray source around the bag for cross-sectional slices. Although quite accurate in terms of P_d, the P_{fa} under operational conditions can unfortunately be up to 0.3 in airports, affording plenty of opportunity for interpretation by operators.

A variety of other machines currently claim to provide automated detection of explosives. For example, London's Heathrow Airport relies upon x-ray machines to automatically select bags with potentially explosive contents. These bags are then subjected to up to four additional increasingly exacting examinations that eventually include screener intervention. However, if a bag with a bomb is missed by the automated level 1 x-ray inspection (all such machines have a $P_d < 1.0$), the bag goes on the plane. In contrast, the TSA's certification standard for the automated detection of explosives would not be met by such level-1 machines. A U.S. certified CT machine does not appear until level-3 at Heathrow.

Figure 3 lists various approaches to inspecting checked bags and other cargo compartment items for explosives. These have been previously mentioned, with the exception of some of those listed under nuclear analysis. These nuclear techniques use highly penetrating radiation as probes and induce penetrating radiation signatures. With TNA, luggage moves through thermal neutrons whose capture in nitrogen results in a high-energy gamma ray. Fast neutron analysis (FNA) improves on TNA by using more energetic neutrons. Impinging upon carbon, oxygen, and nitrogen nuclei, the nucleus scatters the neutron and is placed into an excited state. When the nucleus de-excites, a specific energy gamma ray is released that is characteristic of the target nucleus. The pulsed fast neutron analysis (PFNA) variant of FNA is based on the same concept but uses lower energy neutrons as a probe. Screening applications for TNA, FNA, and PFNA have been limited due in part to cost, size, and slow throughput but are occasionally used in R&D projects.

In addition to the checked bags included in Figure 3, 2.8 million tons of cargo were estimated to be conveyed on passenger planes in 2003. Less than 5 percent of this cargo is physically screened. The primary approach to ensuring security is compliance with the known-shipper program that allows shippers with established histories with carriers or freight forwarders to ship on planes. However, the Department of Transportation (DOT) inspector general (IG) has expressed concern with weaknesses in this program (GAO 2003a). A variation on the conventional use of canines to walk past baggage searching for explosives is that advocated by International Consultants on Targeted Security. Their RASCargo (remote air sampling for canine olfaction) system draws an air sample from the container being examined and then presents the sample to a specially trained dog.

Checked Item Screening for Explosives

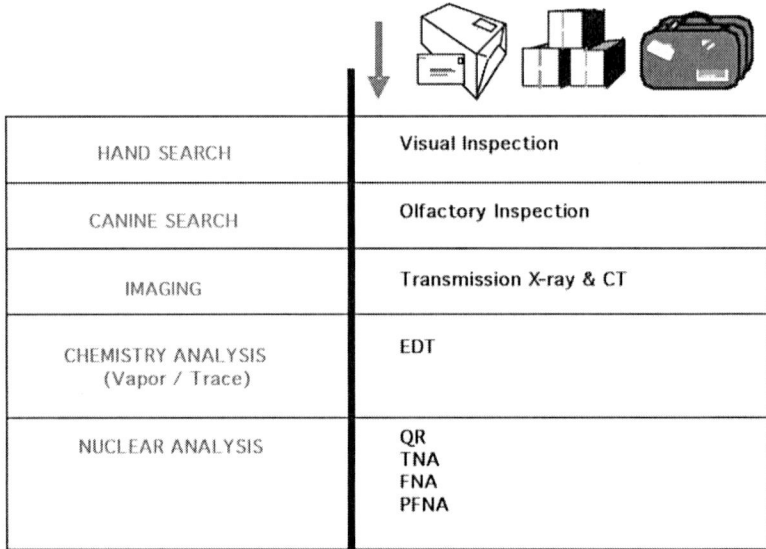

HAND SEARCH	Visual Inspection
CANINE SEARCH	Olfactory Inspection
IMAGING	Transmission X-ray & CT
CHEMISTRY ANALYSIS (Vapor / Trace)	EDT
NUCLEAR ANALYSIS	QR TNA FNA PFNA

Figure 3

Various approaches being used to examine items in the cargo hold

Figure 1 emphasizes the search for explosives and weapons at the check-point, whereas Figure 3 emphasizes the search for explosives in checked bags/cargo/mail. There are, of course, a variety of other potential threats, such as chemical/biological/hazardous (HAZMAT) agents, as well as radiological/nuclear materials. Many types of detectors are available for these threats, such as ASD Biosystems's Biohunter, Smiths Detection HazMatID, and BAE Systems ChemSentry 150C. A difficulty with screening for such threats is that these threats usually cannot be detected until released, presumably after passing through screening. Radiation is much more difficult to contain and, therefore, more easily detected. A variety of detectors are available, such as Ludlum's alpha/beta/gamma detector, and portal detectors also exist, including one for people offered by Canberra.

▶▶ Screeners

Human factors include selection, training, performance certification and assessment, job design, task allocation and workload management, motivation and incentive management, system design and procedures, human interactions with computers and other equipment, perception and attitudes, errors and other behavioral lapses, and health/safety. Prior to the Pan Am 103 tragedy, the human component and contribution to security screening was not adequately considered in terms of its impact on overall system effectiveness. The Pan Am catastrophic event led to the United States Aviation Security Improvement Act of June 1991, which established the FAA's Aviation Security Human Factors Program. Consideration of human factors has the potential to optimize the human contribution within present and advanced technology security systems by accommodating operator constraints and capabilities.

Up to this time, the unstated screening goal was to field ever increasingly complex equipment to eliminate dependence on operators. Technology was typically viewed as the panacea to human performance shortfalls. The inherent flaw in this approach is that eliminating operators requires a technology with 100 percent detection and 0 percent false alarms. This 'holy grail' of security technology is more than a few years away. Therefore, eliminating people from security screening is currently unfeasible, unrealistic, and possibly even unachievable. Paradoxically, new equipment such as EDS, designed to replace operators, has instead increased the mental and workload demands on the operator. As a result, human performance has been made even more central and critical to the overall success of the security system.

Screener Selection

Although significant performance improvement on IED detection had been reported after providing a very limited amount of training, the absolute change was small. It was, therefore, recommended that a screener selection test battery be developed to decrease dependence on training alone by identifying the best-qualified candidates (Fobes and Lofaro 1994). Assessing the potential to improve performance by employing selection tests to assess the aptitude of prospective screener candidates was then initiated. A theoretical framework to identify relevant selection tests was reported (Neiderman and Fobes 1997). In particular, selection tests were chosen in relation to the cognitive processes and strategies of x-ray screening identified in the model presented. The model was developed in an information processing framework that described acquiring and manipulating information that required decisions by the screener. The model described five broad, largely sequential stages of processing: image generation, pattern integration, object recognition, classification, and decision-making. Image generation involved the perceptual identification of basic visual features and the visual search for feature conjunctions. Pattern integration dealt with perceptual organization of visual features and conjunctions to create percepts of coherent forms and objects. Object recognition matched the visual pattern of an object to internal representations of objects and determined the kind of object corresponding to the image. Classification grouped objects into categories of threat, possible threat, and no threat. Decision-making evaluated evidence and generated an appropriate response by using algorithms and heuristics. Two related processes operate on and interact with these stages of processing—attention and vigilance. Attention is the executive control of processing, determining the mental resources devoted to each cognitive process. Vigilance is the maintenance of attention and arousal, a particular problem with infrequently appearing threats such as IEDs.

Selection tests sought were standardized psychometric instruments to measure specific aptitudes and abilities related to cognitive functioning and job performance. The cognitive model provided a framework for identifying potentially useful tests for selection of x-ray screeners. Fifty-one selection tests were described in terms of purpose, procedure, and items and compared to evaluative criteria for a useful test of x-ray screening ability. Each test was reviewed in terms of its procedure and relation to the criteria for a useful selection test. The report recommended a battery of 14 tests for operational evaluation and validation as selection tests. These 14 meet the preliminary criteria of validity, reliability, and practicality likely to prove useful predictors of performance in x-ray screening.

Six perceptual and cognitive tests were eventually tested at 18 major U.S. airports (Neiderman and Fobes 1998, 1999). The ultimate screener aptitude battery (SAB) should predict screeners' performance with both computer-based training (CBT) (next section) and threat detection performance as measured by TIP. To ensure that the SAB would be unbiased, fair, valid, and reliable, components were analyzed for (1) score distribution, (2) item difficulty, (3) item discrimination, (4) test reliability, (5) criterion-related validity, and (6) test fairness. Component tests included the hidden figures test, hidden patterns test, spatial memory test, spatial relations test, visual closure test, and the visual discernment test. The hidden figures test, hidden patterns test, and spatial relations test correlated with CBT performance or TIP detection data, whereas others did not. The hidden figures test and spatial relations test showed adverse impact for minorities and women. While reliability, validity, and fairness concerns were associated with individual tests at this stage, their limitations may be overcome by a future test battery constructed from revisions of the particular tests found to predict learning and training transfer. The strategy for achieving a valid, reliable, and fair selection test battery is an iterative process of fielding the test, revising the test, and fielding the revision. This work needs to be completed.

Pilot work has also been conducted on developing a selection test for CT operators (Cormier and Fobes 1997a, 1997b). Preliminary findings were reported in June 1997 at the point of transferring this project to the FAA's Security Equipment Integrated Product Team created to deploy and operationally evaluate the CTX 5000. The prototype CTX 5000 selection test battery

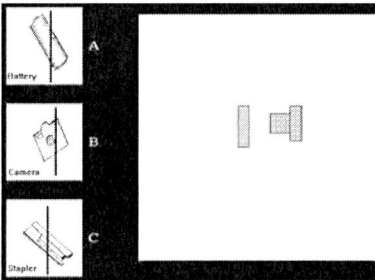

Figure 4
Object identification test. The figure on the right is a representative CT section taken from one of the object choices at the left, at the level of the vertical line. (C is correct.)

consisted of three subtests based on the miniature job sample. These subtests included an object identification test (see Figure 4), where the task was to choose the x-ray image that corresponded to the single CT slice presented; a cross-section identification test (see Figure 5), where the task was to choose the single CT slice that corresponds to a single x-ray image; and an object identification from multiple slices test (see Figure 6), where screeners must choose the x-ray image that corresponded to a set of three CT slices. Each of these subtests consisted of 10 multiple choice items. This project was not initially pursued by the FAA.

Screener Training

Aviation screener training programs were initially limited to the Air Transport Association's 12-hour offering featuring a lecture and slide presentation quite superficial in coverage. Computer-based training began being suggested as a more appropriate vehicle (Fobes and Lofaro 1994; Fobes et al. 1995b), and the FAA eventually approved Safe Passage's CBT in 1997. The FAA deployed the Vocation Station Safe Passage classroom training system for airport demonstration testing (Fobes et al. 1995b). The Safe Passage CBT featured instruction modules for security checkpoint procedures, and a trainee's performance was evaluated with a short test following each module, a final 50-item multiple

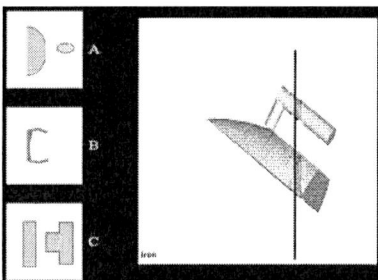

Figure 5
Cross-section identification test. One of the choices on the left was a slice from the object seen in profile on the right, at the level of the vertical line. (A is correct.)

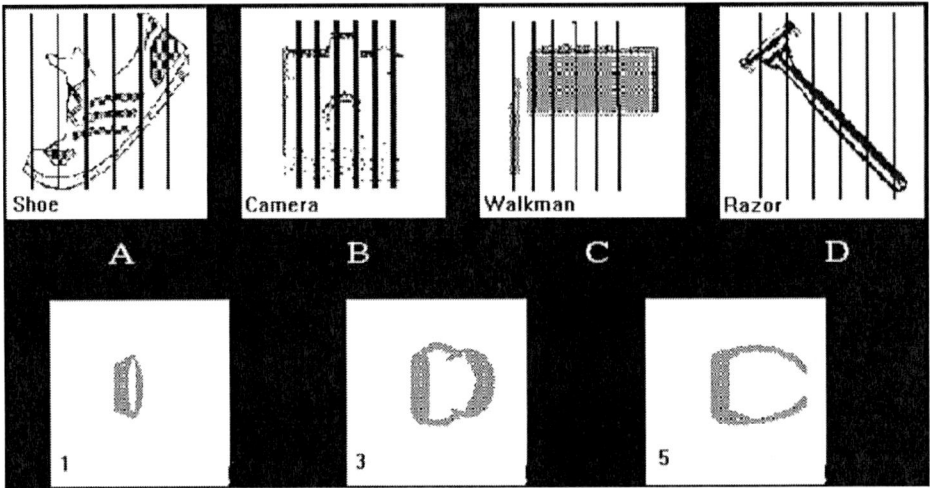

Figure 6

Object identification from multiple slices test. Three CT slices are shown at the bottom. They come from one of the object choices at the top. Slices correspond to parts of the objects between the first and second; third and fourth; and fifth and sixth vertical lines. (A is correct.)

choice test, and a 50-item image interpretation test consisting of judging whether x-ray images contained threats. Performance was examined with a sample of 8,366 CBT tests from 691 screeners at three sites (ATL, DFW, and Seattle-Tacoma International Airport [SEA]) (Fobes and Neiderman 1999). Training mastery testing included examining item readability (content quality and readability), test reliability (test-retest reliability and internal validity), test fairness (unbiased against specific population groups as defined by the Equal Employment Opportunity Act under Title VII of the Civil Rights Act of 1964), and test validity (CBT scores versus TIP performance).

Examination of question quality issues indicated that the overall findings evidenced no effects of item readability (i.e., Flesch-Kincaid Readability (FKR) Index, FKR grade level, grammar, semantic clarity, or wording) on overall performance. Test reliability was characterized as low but was mitigated by the fact that screeners could take tests repeatedly. Adverse impact was found with Asians and Hispanics for CBT completion rates. Adverse impact was also found with African Americans and Hispanics, when initial CBT test scores were examined. However, no adverse impact was found on first-time content and image test scores. The implications of these results are perhaps clearest with the African American group. Average test scores for African Americans on first taking any unit test were lower than those for whites. However, as a group, African Americans persisted in the CBT, repeating unit tests rather than quitting. As a result, African Americans finished the CBT at the same rate as whites and showed no differences in performance by the time the content and image tests were taken. Test validity was indicated by the correlation of both CBT content scores and CBT image test scores with TIP hit rate.

Record keeping and reporting functions are additional advantages of CBT, whether being used for initial, remedial, or recurrent training. The Safe Passage CBT system produced reports for the screener training log, screener training performance comparisons report, individual screener training performance summary report, content test item performance report, and image identification test item performance report. The screener training log provided a high-level overview of the trainees and test results. This report was useful to quickly ascertain the name, screener identification (ID) number, data, and training status of trainees at the airport. The screener training performance comparisons provided additional detail about trainee performance. This report was useful to compare trainee performance scores on each learning module unit test and final examination. The individual screener training performance summary provided a greater level of detail about an individual trainee's performance. This report was useful to examine how well a trainee performed on each test by evaluating each test item that was answered incorrectly. The content test item performance provided detail about each unit test and content final exam question that was presented. The image identification test item performance provided detail about each image interpretation final exam question presented (Neiderman, Klock, and Fobes 1998).

After taking over, TSA increased classroom instruction for screeners from 12 to 40 hours followed by 60 hours of on-the-job training (OJT). Of course, the amount of time being instructed under either process is not a measure of what has been learned or performance effectiveness. A screener readiness test (SRT) project was undertaken to identify the point at which trainees were ready to advance to OJT. The goal was a computerized SRT with the following characteristics: job-relevant, performance-oriented testing of trainees' knowledge; accurate sampling of the entire range of screeners' job functions and roles; and providing a large pool of test questions to minimize the likelihood of cheating. Other requirements for the SRT included psychometric qualities of internal and criterion-related validity, fairness, and user friendliness with minimal demands for administration. The instructional content needed for an SRT, with a job task analysis (JTA) decomposing screening into its component tasks, was described (Fobes and Neiderman 1997). The JTA provided detailed information for determining the knowledge, skills, and abilities (KSA) necessary. The knowledge to be obtained during initial training was represented by four categories. The section on background, responsibilities, and operations for passenger screening included background information, allocation of responsibilities, and overall operations. Identifying the threat consisted of knowledge of threats, their physical appearance, and associated x-ray image appearance. Procedures for passenger screening included general procedures, magnetometer procedures, hand-wanding procedures, pat-down search, x-ray procedures, and bag check. Atypical passengers and special situations consisted of general, equipment malfunction, screening procedures, and weapons.

The skill and ability elements are primarily obtained and honed in the ensuing OJT phase of training, as opposed to knowledge. Behavioral objectives were derived from the KSAs to describe behaviors a trainee must exhibit to perform a task effectively. Objectives included criteria for successful job performance and indicated the conditions under which that performance must be exhibited. Instructional content outlines then provided a vehicle for an instructional development team, customer, and subject matter experts to examine and verify the course content to meet the behavioral objectives. The use of storyboards was described to represent the training interface for computer programmers and multimedia artists to develop lesson modules around instructional objectives. The development of the test item pool to be evaluated for reliability, validity, and adverse impact was described (Fobes, Neiderman, and Klock 1999). X-ray images to be evaluated for the presence of threat articles were developed based on the dimensions of type of threat article (weapon, explosive [five types of bomb with four types of explosives], no threat), type of carry-on item (backpack, briefcase, garment bag, gym bag, overnight case, personal computer carrying case), and level of clutter (two levels).

The prototype set of SRT multiple choice questions (264 questions) and x-ray images of threat and nonthreat bags (1,363 images) was presented to 349 screeners included so that diverse ethnic and gender groups were represented (Fobes and Neiderman 2000). Testing was conducted at JFK, MCO, SFO, and Washington Dulles International Airport (IAD). The reliability of each verbal test module was calculated using coefficient alpha (α) that ranged from $\alpha = 0.69$ to $\alpha = 0.89$ (average $\alpha = 0.84$). Reliability was also calculated for each of the image test modules and ranged from $\alpha = 0.36$ to $\alpha = 0.90$ (average α of 0.76). The SRT item sets had a sufficient range of difficulty to assure score distributions approximately normal with reasonable variance. Item difficulty was relatively uniform across different subject areas. Item intercorrelations were sufficient to produce test modules of moderate length with good internal validity. The test suggested to be fielded for determining whether screeners have been adequately trained consisted of 40 written items (for a maximum of 45 seconds each) and 50 image items (for a maximum of 15 seconds each). The maximum total test time was 42.5 minutes plus instructions. This particular test structure should yield an extremely reliable ($\alpha = .91$) test of the knowledge of newly trained hires. Most importantly, the proposed composite scoring system does not result in an adverse impact on racial and gender groups.

Preliminary work was done to develop a standardized assessment device for OJT performance in a four-phase effort (Monichetti et al. 1998). The project was to (1) describe current OJT practices and curriculum; (2) develop training objectives for required skills and abilities enabling checkpoint tasks; (3) develop a mastery test for OJT skills and abilities; and (4) evaluate and validate the OJT completion test. Current OJT was observed at three airports and found to differ, reinforcing the need for a standardized set of instructional objectives. Observed OJT programs were compared with the screener JTA and KSAs (Fobes and Neiderman 1997). Researchers suggested that accomplishing behavioral objectives could potentially be measured by performance-based testing rather than by the mere passage of OJT time (Monichetti et al. 1999a, 1999b).

Development of an OJT mastery test was proposed to ensure that screener candidates had acquired all of the checkpoint operations skills and abilities as a complement to the knowledge acquired during initial screener training. In addition to x-ray image interpretation, the OJT mastery test would assess candidate screener's skills and abilities to perform magnetometer operations, hand wanding, full and partial pat-down searches, trace detection equipment operations, and carry-on baggage inspections. Screener candidates would be required to perform procedural events via a computer. For example, during the hand-wanding exercise, individuals must demonstrate the correct actions and sequence using a computer mouse and digital 'mannequin.' In addition, they must hold the hand wand in the proper orientation. The final phase was to test, evaluate, and validate the OJT mastery test. This assessment device for determining whether a screener has successfully completed OJT has not been pursued.

▶▶ Screening Systems

Screening equipment needs to be integrated into a security screening system through accommodation of the screener-equipment interaction. There is relatively little accurate equipment automation, which leaves operators substantially involved in the decision-making process. But which screeners, trained by what process, with system performance monitored through what procedure? Prior to passage of the Aviation and Transportation Security Act in 2001, the airlines were responsible for preventing people from bringing threat articles onboard commercial airliners. With few exceptions, the airlines contracted out the work, and security screening was undertaken by individuals meeting minimal qualifications and willing to work for minimum wage. Employee turnover was high, with employees sometimes able to earn more by switching to working in fast-food restaurants at the same airport. Mandatory training was not particularly effective or encompassing. Performance monitoring consisted of a few FAA test articles carried to the checkpoint by frequently recognized compliance inspectors. Since taking over, TSA deployed millions of dollars worth of equipment and more than 50,000 federal screeners at over 440 commercial airports. However, TSA collects little information regarding performance (GAO 2003a). Its primary source of information on threat article detection is covert testing conducted by TSA's Office of Internal Affairs and Program Review. As of November 2003, it had conducted only 733 covert tests at 92 airports, affecting about 1 percent of TSA's screeners (GAO 2003b).

Performance Monitoring

Threat Image Projection consists of an x-ray or CT image of a threat article electronically embedded within the x-ray or CT image of baggage being examined. The screener indicates whether or not a potential threat is noticed, and the screener is given feedback on whether or not a TIP image was present. Done this way, the process is called a fictional threat image (FTI). A different process is used to compare performance across screeners to ensure they are tested under comparable difficulty levels. For example, an FTI within the image of a sparsely cluttered bag might be easier to detect than one in the image of a highly cluttered bag. A combined threat image (CTI) consists of an image of an entire bag containing a threat article. Such testing gives the same problem to each screener. The CTI process requires that the screener cannot see the baggage stream prior to seeing the image displayed. This prevents detections based solely on knowing what the bags' exteriors look like. Only CTIs can be used on CT machines because CTs have the capability to provide the operator with a display of additional requested views or slices around the potential threat object. If an FTI were presented, there would be no information available to the machine to display the requested information revealing that the image was a TIP. A major advantage of TIP is that an IED, in both FTIs and CTIs, can be based on an actual explosive rather than involving an explosive simulant. TSA turned off the TIP feature immediately following the September 11 terrorist attacks, and it was not expected to be fully operational on over 1,770 x-ray machines until April 2004 at the soonest (Berrick 2004). The GAO (2005) reports that TIP is not used for checked bag screening and that, where TIP is activated, TIP is not used as a "formal indicator of screener performance" (Rabkin 2004).

The functional requirements for TIP include naming conventions for threats, data report capabilities, and acceptance test procedures as well as operational and technical criteria (Neiderman et al. 1997). An acceptance test plan was developed to determine the acceptability of individual x-ray machine models (Barrientos et al. 1999) and many machines have been tested (e.g., Barrientos et al. 2000b). A TIP capability is incorporated within most x-ray machines made this decade and has been demonstrated with CT machines (Cormier and Fobes 2000). Using TIP can accomplish a variety of goals for optimizing system detection performance, including enhancing screener training and vigilance and monitoring system performance under realistic operational conditions. TIP received a Technology Innovation Award from *Aviation Week & Space Technology* in 2000 and has a patent (Neiderman and Fobes 2005). For security screening, TIP is the greatest thing since the proverbial sliced bread. Threat image projection can make up for insufficient training or experience by displaying particular threats to be detected. Threats can include emerging threats, such as the credit-card-sized shotgun being sold by a dealer in Minneapolis, that are not in current training programs and would rarely if ever actually be seen by screeners.

A critical factor in screening is the very low expectation of finding a threat other than weapons such as guns or knives. Probably no domestic screener has ever found a real IED, with the possible exception of Charles Dreyling Jr.'s aborted attempt to bring pipe bomb components through security at Will Rogers World Airport in Oklahoma (Airport Business, 2005). This student studying airport management at the University of Oklahoma, and holder of a commercial pilot license, had a carryon with a carbon dioxide cartridge filled with gunpowder and batteries he claimed were for "entertainment value." This low expectation can result in decreased vigilance, and TIP counters this tendency by presenting targets to be detected. The optimal ratio of TIP presentation was reported in a study that presented 6,582 TIP images to certified screeners at Seattle-Tacoma International Airport (SEA). Across ranges from 1/25 to 1/300, the highest P_d accompanied a presentation rate of one TIP per 25 bags (Barrientos 2004). Performance monitoring is perhaps the most desirable feature of TIP. For each TIP presentation, a performance data base contains information on the exact threat article and its category, individual screener, screening company (sites with other than TSA screeners), airport, and airport checkpoint (Fobes and Klock 1998). This basic information can be organized within a variety of reports. For example, detection rates for a particular airport can be obtained as a function of TSA or contracted screeners, exact threat, or threat category (Fobes and Monichetti 1998). User guides have been published to assist federal security directors, security managers, and checkpoint security supervisors to operate TIP. User guides cover user administration (e.g., adding users), management of the TIP library and presentation schedule, viewing reports, and downloading reports as appropriate for the type of administrator (e.g., Fobes et al. 1997).

The potentially massive amount of TIP information supports determining whether individual screeners are meeting the minimum acceptable level of performance. While the bar can be set very high for weapons, finding every IED under the constraints existing at checkpoints (e.g., distractions, time limits, fatigue, inadequacies of x-ray machines) is unrealistic. TIP can provide the information required to determine whether a screener meets minimal P_d for each type of threat. Failures to do so identify when to remove a screener from x-ray or CT inspection and what remedial training is required. TIP can generate a tremendous amount of performance information across the country's airports. Both input (add screeners and TIP images) and output (performance measures) initially had to be done by interacting with the embedded computer in each individual machine. Remote access was then provided for by a process developed to network TIP machines (TIPNET). These functional requirements addressed the computer aspects of network software development (e.g., operating systems, software design), architecture (e.g., file servers, hubs), and components (e.g., transmission media) (Barrientos and Fobes 1999).

As mentioned, TIP is also possible for CT machines and was pilot tested with screeners receiving TIP images on a CTX 5000 at SFO. They retained a fairly high P_d and low P_{fa} across the 3-month project, and performance rates were comparable to those for real bags containing simulated IEDs presented in the baggage flow during this period (Cormier and Fobes 1997c).

So, how well do these systems perform? Prior to 9/11, the FAA had overall responsibility for aviation security. Checkpoint screeners failed to detect 13 percent of the threat articles in 1978 and 20 percent in 1987 (Rabkin 2004). Surprisingly, the former DOT IG published screener test scores on IED detection at particular airports. One rarely sees such information in open sources. The IG reported the following P_ds from FAA testing in 1995 for detecting IEDs made from the MBS. The detection rates at U.S. airports were 0.1 (i.e., 10 percent) for LAX, 0.12 for SEA, 0.15 for Miami, 0.22 for Chicago (ORD), 0.23 for Detroit and Boston, 0.28 for Atlanta, 0.30 for SFO, 0.32 for Houston, 0.42 for BWI, 0.45 for Denver, 0.46 for St. Louis and DCA, 0.53 for Dallas-Fort Worth, 0.56 for JFK and IAD, 0.60 for MCO, and 0.64 for Honolulu (Schiavo and Chartrand 1998). In the months after 9/11 (November to February 2003) while the FAA was still responsible for security, its DOT IG conducted 783 tests at airport screening checkpoints. Screeners failed to detect 70 percent of the knives, 30 percent of the guns, and 60 percent of IEDs, for an overall failure rate of 48 percent (Morrison 2002). More recently, the DHS's IG reported that screener performance had not improved since the 2001 terrorist attacks (Miller 2005).

The GAO has issued a variety of reports on screening performance since TSA assumed the responsibility for transportation security (GAO 2003a, 2003b). The TSA conducted a passenger screener performance improvement study in July 2003 and then developed a short-term screening performance improvement plan in October 2003 to address performance shortfalls. This plan had nine action items (Berrick 2004), some of which had been previously suggested in the 1990s as discussed here. For example, one item proposes increased federal security director accountability for screener performance, a topic addressed in the TIP user guide effort (Fobes and Monichetti 1998). Others deal with the need to install TIP and improve ongoing training, as discussed by Neiderman et al. (1997). Another item urges connectivity to checkpoints, a topic addressed by the TIPNET project (Barrientos and Fobes 1999).

The Aviation and Transportation Security Act moved aviation security to the newly created TSA (February 17, 2002) with the exception of a pilot program retaining private contractors (in TSA uniforms) at airports in one of each of the five categories of airports. These locations are SFO (Category-X), Kansas City, Missouri (Category-I); Rochester, New York (Category-II); Jackson, Wyoming (Category-III); and Tupelo, Mississippi (Category-IV) (Rabkin 2004). The intent was to be able to make comparisons between the performance of government versus privately employed screeners. Although this pilot screening project has been criticized for its extremely small number of participating airports (Poole Jr. 2002), the chairman of the House aviation subcommittee said the GAO found statistically significant superior performance by the private screeners in the pilot program (Miller 2005). The switch to government screeners was not necessarily a permanent change, and additional airports can now apply to join TSA's screening partnership program as of November 2004. This program allows airports to request contracting screening out to private companies, with federal oversight. The Reason Foundation estimated that airports interested in opting out of TSA screening would represent at least 25 percent of originating passengers. The reasons for this include a desire to increase screening quality and obtain greater staffing flexibility (Poole Jr. 2002).

Effectiveness and efficiency can be improved by focusing on those travelers who represent the greatest threat. The FAA has used several profiling approaches to identifying such passengers, including a behavioral one to identify potential hijackers. The FAA's computer assisted passenger screening (CAPS) was developed as an interim solution to identifying whether passengers needed additional security attention (Fobes 1996; Dombrowski and Fobes 1997). It would be interesting to know how many of the 9/11 hijackers were selected by CAPS and whether they were given the additional screening required. Recent efforts to expand the data evaluated under the Computer Assisted Passenger Prescreening System II program have stalled and will seemingly be replaced by Secure Flight. Secure Flight testing began in November 2004 and has airlines transmitting passenger names to the government for comparison with terrorist watch lists (Associated Press 2004). Conversely, the TSA is also testing its registered traveler pilot program wherein participants will go through primary screening (perhaps reserved for them) and not be randomly selected for secondary screening. Travelers must volunteer for the program, provide personal information and biometric data (fingerprint and/or iris scan), and be subjected to a background search (TSA 2004).

▶▶ Six Critical Factors for Security Managers

There are many screening operations being conducted in airports, government buildings, schools, and private venues. The number of screeners will increase as additional venues are added, TSA applies security screening to other transportation modes, and airports opt out of using TSA screeners. Considerable effort is being expended in technology development and transfer for security applications. Much of this effort involves detectors, and security managers need to remain familiar with the equipment options available for their particular applications (Critical Factor 1). One source of such information is the DHS Counter-Terrorism Technology Base (Fobes 2002, 2005a, 2005b, 2005c). The technology base tracks R&D, commercial off-the-shelf equipment, and mitigation strategies for terrorism threats involving chemical/biological/HAZMAT agents, explosives, nuclear/radiological material, and weapons/personnel. Equipment is potentially effective and efficient to the extent that it is usable, and pressure should be brought to bear on manufacturers to optimize the operator-machine interface. In the case of security-related equipment, such emphasis could be applied through the Support Anti-Terrorism by Fostering Effective Technologies (SAFETY) Act (DHS 2003). The SAFETY Act encourages development and deployment of antiterrorism technologies by providing qualified sellers with liability protection from lawsuits. Usability criteria should be made part of the qualification process.

The other aspect of the screening system is the operator. Candidates can considerably differ in their capabilities and motivation. Optimizing the human component begins with careful consideration for the selection of potential operators (Critical Factor 2). Most security managers cannot feasibly develop unbiased, fair, valid, and reliable selection tests, so governmental completion of the R&D cycle for their screener selection test (SAB) is in everyone's interest. Once candidates are selected, a training process must be obtained or developed for the candidates (Critical Factor 3). A computer-based approach has been found to be a superior way to train the knowledge aspect of the security KSAs, particularly for those aspects of screening requiring the detection and recognition of threat objects on a visual display. The security director must then determine when training has been successfully accomplished with a measure based on evaluation rather than on the mere passage of time (Critical Factor 4). The government developed an SRT to identify the point at which trainees were ready to advance to OJT. Some method also needs to be used to evaluate whether necessary skills and abilities of the KSAs have been satisfactorily acquired in OJT (Critical Factor 5). The government should complete the R&D cycle for the standardized OJT mastery test. Finally, and most importantly, objective measures of system performance must be routinely obtained and evaluated (Critical Factor 6). Such information can be the basis for knowing when equipment is malfunctioning (or outdated) or operators need remedial training. TIP is an outstanding measure of system performance with both x-ray and CT equipment. TIP also serves an ongoing training function, presenting threat images of concern. Figure 7 presents the relationship between these six factors.

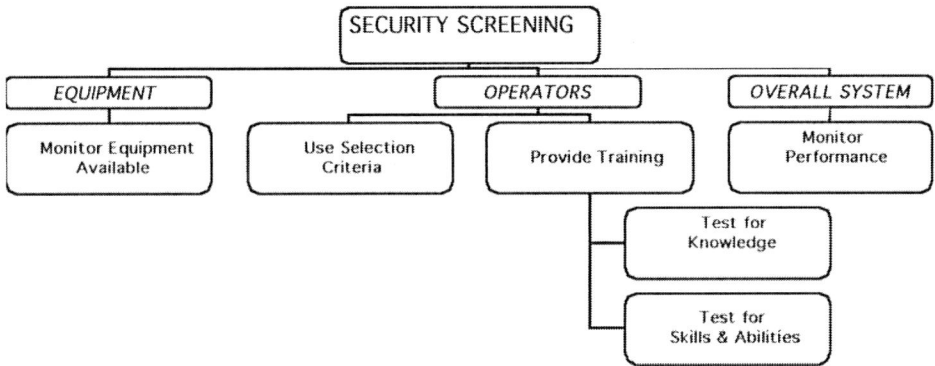

Figure 7
The six critical functions for managers of security screening operations

References

Airport Business. 2005. Man Arrested at Will Rogers World Airport Released on Bail. http://www.airportbusiness.com/article/article.jsp?id=3072&siteSection=5.

Associated Press. 2004. TSA: Tests Going Well for Secure Flight. Voices of September 11, http://www.voicesofsept11.org/security_issues/010605.htm (accessed March 23, 2005).

Barrientos, J. M., J. L. Fobes, K. Goll, B. A. Klock, and E. Neiderman. 2000b. Acceptance Test Report on Threat Image Projection on Vivid Technologies APS Model. Washington, D.C.: Federal Aviation Administration Technical Report DOT/FAA/AR-00/5.

Barrientos, M. 2004. Test and Evaluation Report on the Effect of Threat Image Projection Ratio on Airport Security Screener Performance. Washington, D.C.: Transportation Security Administration Technical Report DOT/TSA-04/xx.

Barrientos, M., and J. L. Fobes. 1999. Functional Requirements for Networking Threat Image Projection. Washington, D.C.: Federal Aviation Administration Technical Report DOT/FAA/AR-99/26.

Barrientos, M., and E. Neiderman. 2000. Test and Evaluation Report for the Evaluation of Screener Assist Technology Using Threat Image Projection. Washington, D.C.: Department of Transportation, Federal Aviation Administration, unpublished report.

Barrientos, M., E. Neiderman, J. L. Fobes, B. Klock, and K. Goll. 1999. Acceptance Test Plan for Threat Image Projection on X-Ray Machines. Washington, D.C.: Federal Aviation Administration Technical Report DOT/FAA/AR-99/32.

Barrientos, M., M. Snyder, and M. Dixon. 2000a. Test and Evaluation Report for the Image Scan Holdings Axis-3D X-Ray Machine. Washington, D.C.: Federal Aviation Administration Technical Report DOT/FAA/AR-00/65.

Berrick, C. 2004. Aviation Security: Challenges Exist in Stabilizing and Enhancing Passenger and Baggage Screening Operations. Washington, D.C.: General Accounting Office Report GAO-04-440T.

Collins, S. 2002. The Manual of Prohibited and Concealable Weapons. Surrey: Intersec Publishing Ltd.

Cormier, S., and J. L. Fobes. 1997a. Test and Evaluation Plan for Selection Tests for CTX 5000 Screeners. Washington, D.C.: Federal Aviation Administration Technical Report DOT/FAA/AR-97/51.

Cormier, S., and J. L. Fobes. 1997b. CTX 5000 SPEARS Status at Transition from Research and Development to Deployment. Washington, D.C.: Federal Aviation Administration Technical Report DOT/FAA/AR-97/70.

Cormier, S., and J. L. Fobes. 1997c. Threat Image Projection Pilot Testing for the CTX 5000. Washington, D.C.: Federal Aviation Administration Technical Report DOT/FAA/AR-97/28.

Cormier, S., and J. L. Fobes. 2000. CTX 5000 SPEARS Status at Transition from Research and Development to Deployment. Washington, D.C.: Federal Aviation Administration Technical Report DOT/FAA/AR-97/70.

Department of Homeland Security. 2003. Fact Sheet: SAFETY Act, Partnering with American Entrepreneurs in Developing New Technologies to Protect the Homeland. Department of Homeland Security, http://www.dhs.gov/dhspublic/display?theme=43&content=3727&print=true (accessed March 23, 2005).

Dixon, M. 2003. CTX 9000 DSi Usability Test and Evaluation Report. Washington, D.C. Transportation Security Administration Technical Report. DOT/TSA/AR-03/xx.

Dombrowski, J., and J. L. Fobes. 1997. Test and Evaluation Report for Computer Assisted Passenger Screening (CAPS). Washington, D.C.: Federal Aviation Administration Technical Report DOT/FAA/AR-97/89.

Federal Aviation Administration. 2002. Federal Aviation Administration Historical Chronology 1926–1996. Federal Aviation Administration, http://www.faa.gov/docs/b-chron.doc (accessed March 25, 2005).

Fobes, J. L. 1996. Computer Assisted Passenger Screening (CAPS). Washington, D.C.: Federal Aviation Administration Technical Report DOT/FAA/AR-96/38.

Fobes, J. L. 2002. Contents Guide to the DOT/FAA's Counter-Terrorism Technology Base (August 00–August 01). Washington, D.C.: Federal Aviation Administration Technical Report DOT/FAA/AR-02/10.

Fobes, J. L. 2005a. Annual Review of Additions to the DOT/TSA Counter-Terrorism Technology Base (August 2001–August 2002). Washington, D.C.: Department of Homeland Security Technical Report DHS/TSA/TSL-05/6.

Fobes, J. L. 2005b. Annual Review of Additions to the DOT/TSA Counter-Terrorism Technology Base (August 2003–August 2004). Washington, D.C.: Department of Homeland Security Technical Report DHS/TSA/TSL-05/8.

Fobes, J. L. 2005c. Annual Review of Additions to the DOT/TSA Counter-Terrorism Technology Base (August 2002–August 2003). Washington, D.C.: Department of Homeland Security Technical Report DHS/TSA/TSL-05/25.

Fobes, J. L., and J. M. Barrientos. 1997. Test and Evaluation Report for Alarm Resolution with X-Ray Screener Assist Technologies. Washington, D.C.: Federal Aviation Administration Technical Report DOT/FAA/AR-97/59.

Fobes, J. L., and B. Klock. 1998. Threat Image Projection Data Base Users' Manual. Washington, D.C.: Federal Aviation Administration Technical Report DOT/FAA/AR-98/78.

Fobes, J. L., and R. Lofaro. 1994. Test and Evaluation Report for Improvised Explosive Device Detection System (IEDDS). Washington, D.C.: Federal Aviation Administration Technical Report DOT/FAA/CT-94/112.

Fobes, J. L., and S. Monichetti. 1998. Threat Image Projection Performance Reports. Washington, D.C.: Federal Aviation Administration Technical Report DOT/FAA/AR-98/77.

Fobes, J. L., and S. B. Monichetti. 1999. Nuclear Quadrupole Resonance and X-Ray Screener Assist: Test and Evaluation Report. Washington, D.C. Federal Aviation Administration Technical Report DOT/FAA/CT-99/90.

Fobes, J. L., and E. Neiderman. 1997. The Training Development Process for Aviation Screeners. Washington, D.C.: Federal Aviation Administration Technical Report DOT/FAA/CT-97/46.

Fobes, J. L., and E. Neiderman. 1999. Validating the Computer-Based Training Process for Aviation Security Screeners. Washington, D.C.: Federal Aviation Administration Technical Report DOT/FAA/AR-99/40.

Fobes, J. L., and E. Neiderman. 2000. Screener Readiness Test Validation. Washington, D.C.: Federal Aviation Administration Technical Report DOT/FAA/CT-00/01.

Fobes, J. L., D. M. McAnulty, and B. A. Klock. 1995a. Screening with Hand-Held Metal Detectors. Washington, D.C.: Federal Aviation Administration Technical Report DOT/FAA/CT-95/49.

Fobes, J. L., E. Neiderman, and B. Klock. 1999. Screener Readiness Test Items. Washington, D.C.: Federal Aviation Administration Technical Report DOT/FAA/AR-99/1.

Fobes, J. L., E. Neiderman, B. Klock, and M. Barrientos. 1997. Threat Image Projection User Guides for Federal Security Managers, Security Company Managers, and Checkpoint Security Supervisors Using EG&G Astrophysics Linescan X-Ray Machines. Washington, D.C.: Federal Aviation Administration Technical Report DOT/FAA/AR-97/80.

Fobes, J. L., M. McAnulty, B. Klock, and M. Barrientos. 1995b. Test and Evaluation Report for Screener Proficiency Evaluation and Reporting System (SPEARS) Computer-Based Training. Washington, D.C.: Federal Aviation Administration Technical Report DOT/FAA/CT-95/43.

Government Accountability Office (GAO). 2003a. Airport Passenger Screening: Preliminary Observations on Progress Made and Challenges Remaining. Washington, D.C.: GAO-03-1173.

Government Accountability Office (GAO). 2003b. Aviation Security: Efforts to Measure Effectiveness and Address Challenges. Washington, D.C.: GAO-04-232T.

Government Accountability Office. 2004. Transportation Security R&D: TSA and DHS Are Researching and Developing Technologies, but Need to Improve R&D Management. Washington, D.C.: GAO-04-890.

Government Accountability Office. 2005. Screener Training and Performance Measurement Strengthened, but More Work Remains. Washington, D.C.: GAO-05-457.

Homeland Defense Journal Online. 2004a. TSA to Conduct Pilot of Cutting-Edge Backscatter Technology. Homeland Defense Journal Online, http://www.homelanddefensejournal.com/scoop_tsa_ferry.htm (accessed March 18, 2005).

Homeland Defense Journal Online. 2004b. TSA Deploys Explosives Trace Portal to New York's JFK International Airport. Homeland Defense Journal Online, http://www.homelanddefensejournal.com/scoop_tsa_ny.htm (accessed March 22, 2005).

Klock, B. 2002. Final Report: Interface and Usability for the i-Portal 100. Washington, D.C.: Transportation Security Administration Technical Report DOT/TSA/AR-02/xx.

Klock, B., W. Maguire, and M. Snyder. 2004. Test and Evaluation Report of the Isorad SDS-400S Fluoroscopic X-Ray System. Washington, D.C.: Transportation Security Administration Technical Report DHS/TSA/TSL-04/xx.

Miller, L. 2005. Report: Private Airport Screeners Outdo Public. *Airport Business*. http://www.airportbusiness.com/article/article.jsp?siteSection=1&id=1852 (accessed June 21, 2005).

Monichetti, S., E. Neiderman, and J. L. Fobes. 1998. Development and Validation Plan for Screener On-The-Job Training Assessment. Washington, D.C.: Transportation Security Administration Technical Report DOT/FAA/AR-98/49.

Monichetti, S., J. L. Fobes, and E. Neiderman. 1999a. Aviation Screener On-The-Job Training Assessment: Content Outlines.Washington, D.C.: Transportation Security Administration. DOT/FAA/AR-99/36.

Monichetti, S., J. L. Fobes, and E. Neiderman. 1999b. Aviation Screener On-The-Job Training Assessment: Knowledge, Skills, and Abilities. Washington, D.C.: Transportation Security Administration. DOT/FAA/AR-99/92.

Morrison, B. 2002. Weapons Slip Past Airport Security. *USA Today*, http://www.usatoday.com/news/sept11/2002/03/25/usat-security.htm (accessed March 23, 2005).

Neiderman, E., and J. L. Fobes. 1997. A Cognitive Model of X-Ray Security Screening: Selection Tests to Identify Applicants Possessing Core Aptitudes. Washington, D.C.: Federal Aviation Administration Technical Report DOT/FAA/AR-97/63.

Neiderman, E., and J. L. Fobes. 1998. Development and Validation Plan for a Screener Selection Test Battery. Washington, D.C.: Federal Aviation Administration Technical Report DOT/FAA/AR-98/47.

Neiderman, E., and J. L. Fobes. 1999. Evaluation of the Screener Aptitude Battery Test Items. Washington, D.C.: Federal Aviation Administration Technical Report DOT/FAA/AR-99/23.

Neiderman, E. C., and J. L. Fobes. 2005. Threat Image Projection System, patent US 6,899,540 B1, 31 May 2005.

Neiderman, E. C., B. Klock, and J. L. Fobes. 1998. Computer-Based Training Performance Reports. Washington, D.C.: Federal Aviation Administration Technical Report DOT/FAA/AR-98/14.

Neiderman, E., J. L. Fobes, M. Barrientos, and B. Klock. 1997. Functional Requirements for Threat Image Projection Systems on X-Ray Machines. Washington, D.C.: Federal Aviation Administration Technical Report DOT/FAA/AR-97/67.

Paulter, N. 2001a. Users' Guide for Hand-Held and Walk-Through Metal Detectors. Washington, D.C.: National Institute of Justice Guide 600-00.

Paulter, N. 2001b. Guide to the Technologies of Concealed Weapon and Contraband Imaging and Detection. Washington, D.C.: National Institute of Justice Guide 602-00.

Paulter, N. 2002. The National Institute of Justice Standards for Hand-Held and Walk-Through Metal Detectors Used in Concealed Weapon and Contraband Detection. Washington, D.C.: National Institute of Justice NISTIR 6915.

Paulter, N. 2003a. Walk-Through Metal Detectors for Use in Concealed Weapon and Contraband Detection. Washington, D.C.: National Institute of Justice Standard 0601.02.

Paulter, N. 2003b. Hand-Held Metal Detectors for Use in Concealed Weapon and Contraband Detection. Washington, D.C.: National Institute of Justice Standard 0602.02.

Polski P. 2004. The Value of OT&E In Homeland Security. Transportation Security Administration, Office of Security Technologies, http://www/dtic.mil/ndia/2004test/tues/polski.ppt (accessed March 22, 2005).

Poole, R. Jr. 2002. *Improving Airport Passenger Screening*. Los Angeles: Reason Foundation.

Pryor, R. 2004. Portal Explosive Detection Development. Transportation Security Administration, Office of Security Technologies, http://www.justnet.org/training/nij2004/pryorAM.pdf (accessed March 22, 2005).

Quantum Magnetics. 2002. i-portal 100 Weapons Detection Systems Using Electromagnetic Anomaly Detection Technology. Quantum Magnetics, http://www.qm.com/products/products_body.htm#i-portyal (accessed March 18, 2005).

Rabkin, N. 2004. Aviation Security: Private Screening Contractors Have Little Flexibility to Implement Innovative Approaches. Washington, D.C.: General Accounting Office Report GAO-04-505T.

Rhykerd, C. L., D. W. Hannum, D. W. Murray, and J. E. Parmeter. 1999. Guide for the Selection of Commercial Explosives Detection Systems for Law Enforcement Applications. Washington, D.C.: National Institute of Justice Guide 100-99.

Schiavo, M., and S. Chartrand. 1998. *Flying Blind, Flying Safe*. New York: Avon Books.

Smiths Detection. 2004. Smiths Detection Provides High-Speed Document Scanners for TSA Pilot Program at Four Major U.S. Airports. Smiths Detection, http://www.smithsdetection.com/PressRelease.asp?autonum=35&bhfv=6&bhsh=600&bhsw=800&bhiw=789&bhih=446&bhsp=550287&bhqs=1 (accessed March 25, 2005).

Thompson, Elwyn. 2004. Bottle Screening Devices. Federal Business Opportunities, http://www.fbodaily.com/archive/2004/08-August/27-Aug-2004/FBO-00653219.htm (accessed March 22, 2005).

Transportation Security Administration. 2004. What Is Registered Traveler? Transportation Security Administration, http://www.tsa.gov/interweb/assetlibrary/Factsheet.pdf (accessed March 23, 2005).

U.S. Department of Homeland Security. 2004. TSA Launches New Passenger Rail Security Project. U.S. Department of Homeland Security, Transportation Security Administration, http://www.tsa.gov/public/display?theme=44&content=09000519800a0cda (accessed March 22, 2005).

Webshots. 2005. U.S. Foresaw Terror Threats in 1970s. Associated Press, http://daily.webshots.com/content/ap/current/h63802950.html (accessed March 18, 2005).

Chapter 7

Bioterrorism Training Simulations: Implications for Effective Response

PAUL DRNEVICH, ALOK CHATURVEDI, AND SHAILENDRA MEHTA
Purdue University

▶▶ Introduction

The threat of terrorism today haunts practically every nation in the world, and the perceived threat of terrorism in the United States has increased dramatically this decade. Recent terrorist attacks and evidence of continued potential threats indicate that governmental agencies at the federal, state, and local levels must be prepared to face unique and extreme pressures and must make decisions with implications of extreme magnitude. Terrorism can take many forms, but the most serious threats are posed by terrorists armed with weapons of mass destruction (WMD). In fact, some U.S. government officials have gone so far as to say that it is only a matter of time before such an attack with nuclear, biological, or chemical weapons takes place on domestic U.S. soil (Cheney 2002). Of these threat scenarios, attention has been focused on bioterrorism weapons because, unlike nuclear weapons, they are easy to make and, unlike chemical weapons, they require only small quantities to produce devastating results.

To help counter these potential WMD threats, government agencies are looking at science and technology as a means of augmenting and refining traditional existing military and law enforcement assets. To deal with these unique and challenging new situations, government agencies must develop and practice coordinated response strategies for possible terrorist strikes in the United States, and the nature of the operational pressure, turbidity, and outcome implications make this a unique and challenging operating environment. Toward this end, we developed a simulation training exercise series termed Measured Response (MR) at Purdue University to facilitate response decision making under these conditions. Through the exercise federal, state, and local officials and others representing the Departments of Homeland Security, Health and Human Services, and Transportation were able to practice response skills and engage in stimulating interaction through a simulation training exercise.

The overall collective goal of the government when responding to a terrorist attack is to minimize the effectiveness of the WMD attack in terms of casualties. This goal is somewhat complicated by the need to minimize public panic regarding the attack. Therefore, the government agencies must consider the impact on public sentiment that their decisions may have. Specifically, the wider implications of the attack must be considered in terms of the impact and effects on the whole population. Improving our response and coordination capabilities is, therefore, a persistent challenge to response effectiveness, one that can hopefully be improved through the application of science and technology to these issues.

Clearly, there is a need for an innovative platform that addresses the issues outlined above, as well as for use to develop, test, analyze, and implement public policies and operating procedures for terrorism response. The Measured Response simulation is an example of an environment that permits such research. Toward this end, we first describe the Measured Response exercise series. This is followed by an overview and observations of the 2002 and 2003 Measured Response exercises. We next provide an overview of the simulation system. This is followed by the development of a model. We conclude with a discussion of the results and implications for both practitioners and further academic research.

▶▶ Background on Bioterrorism Response Training

Biological warfare or bioterrorism is defined as the use of pathogens that are disease-causing bacterial and viral agents or biologically derived toxins against humans, animals, or crops (Croddy et al. 2002). One of the main goals of bioterrorism is to disseminate disease-causing bacteria, virus, or fungi to inflict massive casualties on the targeted territory. As many as 17 countries are known to have extensive bioweapons programs, and a further 100 nations are estimated to have the expertise, technology, and resources to establish a bioweapons program (U.S. Army 2001). Beyond state-sponsored programs, the combination of globalization and the ease of information dissemination in this high-tech information age have made it possible for individuals and groups to manufacture their own bioweapons for evil intents.

As a case in point, the Aum Shinrikyo cult in Japan in 1997 scared the world by staging a sarin nerve gas attack, killing 12 and injuring more than a thousand people. Previously, the cult had tried (and failed) several times to disperse biological agents such as anthrax by using fans and sprayers. An even earlier event occurred in 1984 when bioterrorism hit home when members of the Bhagwan Shree Rajneesh cult field tested their homegrown *Salmonella typhimurium* bacteria by pouring it in a salad bar. The objective was to sicken individuals and prevent them from voting in order to swing votes and take control of the rural county commission. Close to 700 people were sickened by the attack (Chyba 2002). Similar threats and hoaxes in the 1990s have further terrified local, state, and federal authorities in the United States. These hoaxes and threats included several bioagents such as *Shigella dysenteriae, Anthracis yersinia*, and ricin. To deal with these potential situations, a number of training exercises have been developed to improve the effectiveness of response to an actual event. These include the Dark Winter, TOPOFF, and Measured Response exercise series (Inglesby et al. 2001; O'Toole and Inglesby 2002).

The Dark Winter Exercise Series

Dark Winter was an exercise conducted in June 2001 by Johns Hopkins University and the Center for Strategic and International Studies in conjunction with the State of Oklahoma to examine how senior policy makers would make decisions in the face of a bioterror attack (O'Toole and Inglesby 2002). The pathogen used in this exercise was the smallpox virus, which, in spite of its eradication in the general population, is stored in two official repositories maintained by the World Health Organization in the United States and Russia. In addition, it is widely assumed that it might be in the possession of several rogue states. The exercise was a tabletop exercise, which distinguished it from two other types of exercise in common use—field exercises, where real-world processes and resources are used, and simulation exercises, where a formal, usually computer, model is used to set the stage. In Dark Winter a script representing the spread of the smallpox was played out in three segments representing a period of two weeks. Over this period, experienced public officials played the roles of the president of the United States, the president's national security advisor, several key cabinet members, the governor of Oklahoma, and others. Several experienced journalists and reporters played media-related roles. The role players had several key decisions to make. Based on the script detailing the spread of the virus and the resulting distribution of people infected and dead, the policy makers had to decide how to allocate the limited vaccination resources available to the government. In addition, issues of confinement, patient care, restrictions on international travel and panic were examined.

The lessons of Dark Winter (as summarized by O'Toole and Inglesby 2002) were as follows. First, leaders were generally unfamiliar with the nature and character of bioterror attacks. As a result, inputs from the medical and public health community were crucial. Next, a shortage of vaccines limited the options. Then, the U.S. health care system lacked surge capacity to deal with mass casualties. Then, priorities of agencies at various levels differed or were even in conflict, and finally, gaining the trust and cooperation of the citizens was essential.

The TOPOFF Exercise Series

The TOPOFF series of exercises, by contrast, are field exercises. They were mandated by the U.S. Congress (the 1996 Dunn-Lugar-Domenici Act) and require the engagement of the top officials of the government, hence the name. The exercises are conducted biennially, and three such exercises have been conducted so far. TOPOFF-1 was conducted between May 20 and May 23, 2000, and featured three separate attacks: Denver, Colorado, which was targeted with the plague; Portsmouth, New Hampshire, which was targeted with a chemical attack; and Washington, D.C., which was targeted with a radiological attack. TOPOFF-2 was held in Chicago, Illinois, and Seattle, Washington, from May 13 to May 16, 2003, and featured a radiological attack and a bioterror attack, respectively. TOPOFF-3 was held from April 4 to April 8, 2005, in Union and Middlesex Counties in New Jersey, where a chemical attack was used, and New London, Connecticut, where a bioterror attack with the plague bacterium was used. Surprisingly, very little public information or analysis about these exercises exists.

A partial account of TOPOFF-1 is provided by Inglesby et al. (2001), and some details of the TOPOFF-2 and TOPOFF-3 exercises are available in media reports (e.g. Kelley 2005). Among the lessons learned from these exercises, we may count the following. First, both effective political leadership, as well as expert inputs, are important to making effective decisions. Second, the absence of real-time information on the spread of casualties was found to be a major issue impeding decision making. Third, the quantity and distribution of resources available (including hospital capacity) was insufficient to deal effectively with such crises. Fourth, it was found that clearly articulated principles that passed both scientific and political tests were critical. Finally, the direct costs of each TOPOFF exercise are estimated to be in the range of $16 million, and the indirect costs in terms of people's time and inconvenience are likely to be even higher. Therefore, "careful consideration" must be given to the "objectives and purposes of future exercises" with a view to enhancing cost effectiveness (Inglesby et al. 2001).

Measured Response Exercise Series

Similar to and concurrent with the Dark Winter and TOPOFF exercise series (Inglesby et al. 2001; O'Toole and Inglesby 2002), we developed a training exercise series, termed Measured Response (MR), at Purdue University to facilitate response decision making for these types of events. However, the Measured Response exercise, by contrast, is a computer simulation exercise and, therefore, different in form and character from the tabletop and field exercise formats used in Dark Winter and TOPOFF. Simulations, especially when built on an extensible platform, once developed, can be reused multiple times and provide a number of advantages. These include the facts that extensible simulation platforms can involve people from around the world who do not need to be assembled in one location, can be rolled out at low cost per participant, can be captured and rerun for the benefit of later observers, can be extended to multiple interacting regions, and can be modified on the fly as necessary to inject unplanned uncertainty and realism. Their main drawback is their relatively high level of upfront development cost and complexity. Given this, they become only viable with reuse and broad application. Realism can also suffer in simplistic simulations, but by making models more extensive and complex, this drawback can also be overcome to some extent. Table 1 summarizes the characteristics of the three kinds of simulations.

Table 1. Comparison of terrorism response training exercises

	Tabletop (e.g., Dark Winter)	Field (e.g., TOPOFF)	Simulation (e.g., Measured Response)
Fixed cost	Low	Very high	High
Variable cost	High	Very high	Low
Realism	Variable	High	Medium
Scope	Broad	Narrow	Very broad
Participation	Limited	Broad	Very broad
Reusability	High	Low	Very high
Ability to modify	High	Low	Very high
Ability to replay	Low	None	Very high

Measured Response (MR) is a synthetic environment that simulates the consequences of a bioterrorism attack on a midsized city. The objectives of the exercise are to develop and analyze policies and operating procedures to manage the public mood, maintain public health, mitigate the risk of contagion, maintain orderly movement of traffic and people, and apprehend perpetrators. The MR exercise enables participants to work on key response skills, which include risk management, prioritization, communications, incident management, cooperation, and management of public mood. Key training objectives of MR include: practicing resource/risk management under an unconventional crisis situation; prioritization, timing, and intensity trade-offs of response decisions and actions; testing emergent communication strategy development and enhancement; dealing with real-time incident management; allocating decision-making among different levels; execution and effort coordination among different agencies and actors; and the management of public mood and expectations.

Measured Response models the rate of transmission as a function of population density, mobility, social structure, and lifestyle using an explicit spatiotemporal model. It uses the movement of individuals and the exposure of susceptible individuals to infected individuals to model the spread of disease. It models the communicability of infections both from host to host (small pox, influenza, and Ebola) and from the environment to the host (anthrax). In addition to the standard epidemiological parameters such as reproductive rates of infection and disease propagation rates among individuals, MR also models the hosts and pathogens via several interrelated processes. These include age-specific susceptibility, infection propagation due to the exposure of wholly susceptible populations to newly infectious population, and population immunity necessary to prevent the epidemi.

Measured Response 2002

In the 2002 Measured Response exercise (MR02) government officials at each level and agency noticed the conflict of interest between elected officials and responders. This was based on role dependency on public mood, which is essential for elected officials due to job security and reelection concerns. Conversely, responders tend to seek prompt action (military quarantine), which, although effective in combating the contagion by containing the outbreak to the local population, negatively affects public mood within and beyond the response area. The government will thus disagree about level and timing of information outflows to the public. Due to this conflict and inaction, the outbreak was not immediately contained in the simulated Ebola outbreak spread. Observations from MR02 were that the government responders needed to take decisive, coordinated, and timely actions. However, response will vary by situation and involve a trade-off between decisiveness/quickness of response and the need for certainty of response. Specifically, from MR02, government officials at each level learned the need for an incident command center, defined roles for different agencies, and reduced decision cycle times.

In MR02 government officials clearly struggled with the communication process, mainly due to a lack of guidelines between and within levels. As such, the importance of a well-defined information flow became evident. It is essential that the right people have the right information at the right time. As discovered through the MR02 simulation, a policy is needed to streamline the flow of information across levels. With such an approach, the diagnosis of the disease can be done more efficiently when different personnel are informed of the events. Based on the findings from MR02, requirements for a coordinated measured response communications strategy should include decisiveness; incidence management and the delegation of authority; decision-making process and the timeliness of action; risk assessment, management, and mitigation; resource allocation across different functions; communication strategy execution and coordination (nongovernment organizations [NGOs] media, division chiefs); and management of public perceptions and expectations.

Measured Response 2003
The 2003 Measured Response exercise (MR03) was the second annual event in the exercise series and featured a scenario simulating a smallpox outbreak. In addition to the standard epidemiological parameters, such as reproductive rates of infection and disease propagation rates among individuals, MR also models the hosts and pathogens via several interrelated processes. These include age-specific susceptibility, infection propagation due to the exposure of wholly susceptible populations to a newly infectious population, and population immunity necessary to prevent the epidemic (Chaturvedi and Mehta 2002; Chaturvedi et al. 2005).

In the exercise, nine teams of three individuals each (representing their actual responsibilities in most cases), played the roles of the departments of homeland security, human and health services, and transportation at the local, state, and federal levels. The participants interacted with the simulation in a war room environment. In the exercise, human players were able to affect the transmission rates through different intervention actions such as quarantine and vaccination. Large-screen displays allowed the participants to obtain a high-level overview, as well as a real-time detailed account of the data generated during the exercise rounds.

The response to the attack in MR03 was based on available real-world options and depended on the decisions of the human actors representing their respective government agencies. The decision makers were tasked with determining the proper course of action to implement across each of five cities in the state to combat the virus outbreak. The scenario began with local government officials learning that a body (a virtual agent) with an unknown cause of death had been discovered in a downtown hotel in the capital city. Evidence indicated that the deceased may have been infected with a virus and possibly came in contact while infectious with many people at a downtown music festival. The local government officials faced several choices, each with ramifications that could determine the success or failure of the potential attack. Key response issues included both response timeliness and resources (if and when to involve state and federal officials), as well as response action and focus (should a quarantine be imposed to isolate the threat and, if so, over how widespread an area?).

In the MR03 simulation, local officials determined that the initial casualty had indeed been involved in a biological attack against the city. State and federal officials were brought in to assist with mitigating the damages and to help in response and containment. Based on supplied real-world operational data, the government officials were allocated six decision choices, with three commonly accepted levels of intensity to apply to the actions. These six response choices were variants of quarantine or vaccination strategies or no response. The level of intensity varied from neutral/no action to mild/moderate action and up to extreme action. The corresponding level of success or failure depends on how the thousands of intelligent agents react to, and are affected by, the government agent's responses in the virtual environment. Thus, if effective quarantine strategies were chosen, and the spread of infection was slowed, then this would be reflected in lower casualties among the artificial agents, as well as a higher score for public mood. Conversely, ineffective measures would have the opposite effect.

When responding to a terrorist attack, the overall collective goal of the government agencies was to minimize the effectiveness of the attack in terms of contagion spread (infection and death rates) and public perception. This goal was also somewhat complicated by the need to maintain acceptable levels of public perception regarding the attack. In this, the government agencies must consider the impact their decisions may have on public sentiment. Specifically, the wider implications of the attack must be considered in terms of the impact and effects on the whole population. For this reason, the public mood was measured to look at outcomes beyond measures of infection and death rates. The public mood was operationalized as the level of happiness of the artificial agents representing the population and was influenced by the response strategy, intensity, and timing. In this construct several factors influence the public mood of the civilian population, simulated by artificial agents, when reacting to a threat in the MR03 simulation. These factors include security, basic necessities, health, mobility and freedom, weather, information level, financial capability, and the global economy (Chaturvedi et al. 2005).

Synthetic Environment for Continuous Experimentation (SECE)

The Measured Response exercises take place in the Synthetic Environment for Continuous Experimentation (SECE) (Chaturvedi and Mehta 2002). SECE uses an explicit spatial and temporal paradigm to develop a synthetic society that mimics the essential demographic, epidemiological, and economic characteristics of the United States.

SECE is developed on the Synthetic Environment for Analysis and Simulations (SEAS) platform (Chaturvedi and Mehta 2002, Chaturvedi et al. 2005). SEAS allows the creation of fully functioning synthetic economies that represent the real economy in many key aspects by combining large numbers of artificial agents with a relatively smaller number of human agents to capture both detail-intensive and strategy-intensive interactions. In MR, several hundred thousand artificial agents can be employed to mimic the behavior of the citizens. Such behavior includes: mobility, the feeling of well-being in terms of security (financial and physical), health; information, and civil liberties. The artificial agents run on a distributed tera-scale grid computing environment comprised of two supercomputers connected by a Gigabit network. We accomplish this by creating hundreds of thousands of artificial agents that have multiple layers of attributes and behaviors representing the position, mobility, infectability, and well-being of the citizens of the United States. Therefore, each artificial agent can represent about 1,000 real people. Further, this virtual geography is similar to the real geography, and the population of artificial agents is tied to different locations accordingly. The demographics of the artificial agents also mimic the demographics of the actual population. The epidemiological characteristics, such as susceptibility to infection by age, gender, and race, broadly reflect the real data. In addition, the well-being of individuals in the model is drawn from well-established paradigms in economics and psychology.

Using this environment, we are able to simulate a biological attack on a synthetic population and allow the human teams to respond. Using carefully constructed algorithms (Chaturvedi et al. 2005) that allow the virtual agents to replicate actual human activity, the attack propagates through the virtual population as it would through our society. Upon the completion of a scenario, the government officials are informed of the ramifications of their decisions, which they then will be able to analyze for impact. From this comparison, the officials are able to learn to improve both their decision-making and response time. The tool allows officials to examine response plans in a risk-free environment, thus enhancing their ability to respond in an actual attack (see Figure 1 for an overview).

The environment describes the background and the contextual structure of the model. It lists the entities and describes the relationships among them. It contains the geography and the physical details of the space, such as the

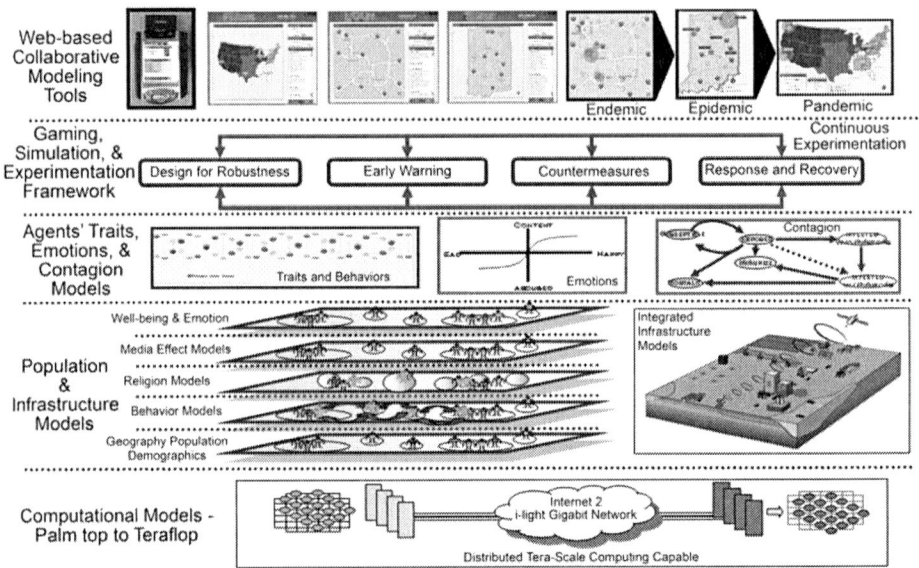

Figure 1

Overview of Measured Response Simulation

road networks, the structures, traffic patterns, and pedestrian dispersion. It also implements the rule sets that guide the interaction of the agents between each other and with the environment. Hundreds of thousands of synthetic, computer-based agents, whose emergent behavior defines the environment, are used as the platform on which human players can engage in strategic decision-making simulations. The environment in this context is a hybrid of microeconomic analysis combined with models from the fields of operations research and management science, epidemiology, and psychology (although a much different, bottom-up kind of simulation than the typical top-down discrete event simulation).

Situated in this simulation environment are agents. An agent typically represents a group of people with similar characteristics and demographics in the simulation. These agents can interact with other agents and with the environment. A distinction is made between artificial agents and human agents. The roles of these agents can be interchanged based on the requirements of the problem domain. The behavior of the human players is not predetermined, and they are free to act as they wish under existing conditions (which might include various capabilities and constraints), which are very clearly described and presented to them in an intuitive and informative way. The behavior of artificial agents is rule based, and the rules may involve randomness and may evolve over time. Each agent's behavior pattern is one of action, reaction, and counteraction. This is elicited in response to the environment, where the agents respond to other agents, as well as to stimuli such as player responses to the bioterror attack. The agents contain autonomous processes that are adaptive and behave like human agents in a narrow domain. In their respective domains, each agent has a well-defined set of responsibilities and authorities so that it can execute its tasks effectively.

Each geographic entity (city, state, country) has a certain productive capacity that includes labor, capital, and natural resources. In the event of a terrorist attack, the production of goods and services is disrupted. This disruption is explicitly modeled using publicly available data. In addition, response to an attack involves directing resources to mitigate the effects of attacks. These include emergency provision of food, clothing, and shelter, as well as medical supplies. Hospital beds have to be allocated and vaccination, quarantine, and isolation programs have to be activated. In addition, demands for goods and services are also directly built into the behavior of artificial agents. Some agents will engage in rather simple rules of behavior (Gode and Sunder 1993; Gode and Sunder 1997), while others will be more complex (Chaturvedi and Mehta 2002).

To focus on our main goal, the computational model of human behavior includes two components, mobility and the feeling of well-being. The mobility model includes normal movements, morning and evening rush hours, event-related movements, and panic fleeing. We also model action tendencies based on research of individuals and crowds. Collective behavior, characterized by specific directional and structural movements, often emerges in crowds gathering and flowing in limited space (Milgrom and Toch 1969). Their organized patterns of behavior should be regarded as the macroscopically structured behavior of a crowd and not an unstructured aggregate of individual pedestrians (Yamori 1998). SECE incorporates, where appropriate, the micro-macro linkages in agent behavior. These are appropriate for modeling when crowds become agitated and violent during attempts by civil authorities to impose order upon the chaos that might ensue upon a bioterror attack.

Finally, to ensure that the attack propagation reflects real-world dynamics, Measured Response employs an explicit spatio-temporal paradigm to model the spread of an epidemic over time and space. In the SECE environment, as in the real world, reproductive rates and propagation vary according to the type of attack. Similarly, variables such as population density, agent mobility, social structure, and way of life interact to determine the propagation of infections. The government officials, or human agents, interact with the system and control propagation via such means as vaccination, treatment, or agent isolation. The options available to the human agents are the same as in real life, and the effectiveness of these interactions is modeled using statistically verified historical information (Rvachev and Longini 1985).

▶▶ Simulation Model for Measured Response

In Measured Response, unmodeled emergent strategic behavior (rooted in prior real-world experience and training) resulted in less than ideal outcomes. Observations from MR02 identified a variety of shortcomings in response coordination with respect to strategy choice, level of intensity, and timing of the response. Our current study explores these issues by modeling the observed decision-making process and examining the predicted implications. The observed behaviors and actions in MR02 and managerial theory were used in conjunction to model the response activity that takes place in real-world situations. From the expected theoretical issues and the observed actions in the MR02 exercise, we propose a model for the outcome effectiveness of a government agency coordinated response as a function of the strategy decision choice, the timing of the strategy implementation, and the intensity of the strategy implementation. This model is depicted in Figure 2.

The components of our coordinated response effectiveness model consist of the strategic response choice, which is moderated by the intensity and timing of the strategy implementation, and the resultant outcome. The strategy parameter refers to the strategy choice utilized by the responders (variants of vaccination and quarantine options). Intensity refers to the level of effort applied to moderate the response strategy (mild or extreme). The timing parameter refers to whether a response is immediately after the attack, after a slight delay, or after an extreme delay. The outcome parameter refers to the relative success or failure of the strategic choice and contingent responses that may consist of a variety of measures (i.e., number infected, number deceased, public mood, etc.).

Response Strategy

Factors of strategy choice involve the management of incidents arising from the simulated bioterror attack in this exercise. This strategic decision-making process involves the proper assessment of the risks from available information and the management and mitigation of the perceived risks. Once the information is processed and the risk is assessed, a response strategy can be

Figure 2

Strategy Choice and Response Effectiveness Decision Model

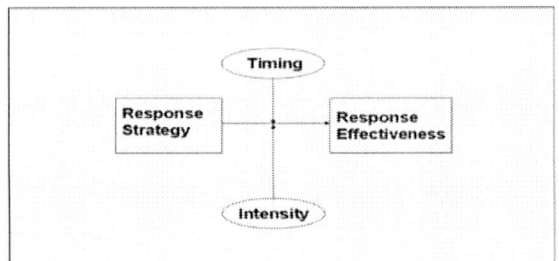

formulated. Strategy choices for dealing with the bioterror attack on a city and its potential spread to other cities involve options of quarantine and vaccination response strategies. We classify these response strategies into categories based upon the degree of restrictiveness: no quarantine, partial quarantine (city block), and total quarantine. The type of quarantine strategy may also affect the mood of the population. For example, commonly accepted federal emergency and military response practices, such as mass quarantine or mass vaccination, may be perceived as overly restrictive by the affected population as well as by the unaffected population (Kaplan et al. 2002). These response choices are by nature more likely to be effective in containing and stopping the virus outbreak but also are likely to negatively affect public mood (Kaplan et al. 2002). Less restrictive strategies include city block and trace vaccinating, as well as city block quarantining. This class of response choices is less likely to negatively affect public mood but also is theoretically less likely to effectively contain the virus outbreak (Kaplan et al. 2002). Restrictiveness of response choice will, therefore, likely be negatively related to public mood but positively related to response effectiveness in terms of containing the virus outbreak.

Response Intensity
After the response choice is selected, the agents must decide upon the level of intensity of response implementation. As stated previously, the effects of the intensity of the response strategy will moderate the outcome effects of the strategy choice. The selected level of intensity may vary and will affect the execution and coordination of the response strategy as well as influence the public mood and the cooperation of nongovernment organizations such as the media. In choosing the response intensity level, the management of public perceptions and expectations is an important consideration and perhaps most directly linked to public mood. As in the private sector, the intensity level of the strategy implementation will also involve resource allocation commitments across different functions (i.e., which assets should respond and how many resources should be committed to the response). The theoretical implications of these hypothesized moderator effects will likely be positively related with the outcome in terms of containment (Kaplan et al. 2002) but negatively related with public mood (Chaturvedi et al. 2005). The level of intensity moderating the response strategy will be positively related to strategy outcome effectiveness in terms of containing the virus outbreak but negatively related to strategy outcome effectiveness with respect to public mood.

Response Timing
The effects of the strategy choice and the implementation intensity decisions must also be coordinated with the timing of the response. The timing of the vaccination or quarantine strategy assumes no action is taken until the attack is

evidenced. The response can therefore only take place immediately after the attack, after a slight delay, or after an extreme delay. Then the appropriate vaccination or quarantine strategy choice is implemented by the players. The implications of this moderator involve the issues of decisiveness and the timeliness of action in the decision-making process. A quick response may be essential for containing the virus outbreak but requires fast action from the decision makers. This involves making decisions with incomplete and uncertain information, where incorrect decisions may result in a lack of containment of the contagion, as well as in negative effects on the public mood. However, a slow or late response may be just as detrimental to the outcomes because failing to act quickly allows the contagion to spread. This further erodes the public mood as casualties mount and fear and panic spread among the affected and infected population. Response timing decisions, therefore, must consider these issues and seek to achieve an optimal balance of outcomes given these trade-offs. A quick response may be more effective at containment but may be improperly targeted or misdirected, resulting in panic and collateral damage. A less timely response, after sufficient situational information has been processed, may be more accurately directed but less effective in rapidly containing the virus outbreak. On the one hand, responding too aggressively, too early, may adversely affect public mood and produce collateral casualties. On the other hand, responding too late may allow the contagion to grow beyond the control of the responders. Given these implications, the trade-off between certainty and timeliness must be balanced by the decision makers. The timing in implementing response strategies will, therefore, be positively related to strategy outcome effectiveness in terms of containing the virus outbreak. Further, the timing in implementing the response strategy will be negatively related to strategy outcome effectiveness with respect to public mood.

▶▶ Results and Observations

Infected, Deceased, and Public Mood

As mentioned earlier, the overall collective goal of the government agencies is to minimize the effectiveness of the attack in terms of contagion diffusion (infection and death rate), but because this goal is somewhat complicated by the need to maintain acceptable levels of public perception regarding the attack, the government agencies must consider the impact of their decisions on public sentiment. These wider implications of the attack are considered in terms of the impact and effects on the whole population through the construction of public mood. Public mood is utilized for control purposes beyond outcome measures of infection and death rates. As discussed earlier in detail, factors that influence public mood of the artificial agents in the measured response environment include security, basic necessities, health, mobility and freedom, weather, information level, financial capability, and the global economy (Chaturvedi and Mehta 2002). The public mood is, therefore, directly influenced by government response choice, intensity, and timing in the same manner as infection and death rate.

The variants of the previously discussed model were run for each of the outcomes (number exposed, infected, death rate, and public mood). The first variant, number exposed, is measured as the total count of the population exposed to the outbreak. The second variant, number infected, is measured as a combination of the total count of infected with symptoms and infected without symptoms following a simulated 40-day response period using strategy, intensity, and timing response combinations to the bioterror attack. The third variant, number deceased, is measured as the total count of deaths following a simulated 40-day response period using strategy, intensity, and timing response combinations to the bioterror attack. The fourth variant measured is public mood, which is measured on a seven-point Likert-type scale during the simulated 40-day response period using strategy, intensity, and timing response combinations to the bioterror attack. This measure produces a relative range score, where higher scores (5 to 7) indicate a more positive public mood and lower scores (1 to 3) indicate a more negative public mood. Further details on the variables are listed in Table 2.

Response Strategy, Intensity, and Timing

The other variables of the model consist of the response strategy, the intensity of the strategy implementation, and the timing of the strategy implementation. The set of participant action response strategies can be divided into three main categories: (1) no human intervention, (2) vaccinating, or (3) quarantining. These strategy choices for dealing with the bioterror attack on the city and its potential spread to other cities involve options of quarantine and vaccination response strategies. The intensity of the strategy implementation is correlated with the strategy choice. The timing of the vaccination or quarantine strategy implementation assumes no action is taken until the attack is evidenced (days 1 to 4). The response implementation can, therefore, only take place immediately after the attack (day 5), after slight delay (days 6 to 7), or after extreme delay (day 8).

The strategy choices for dealing with the bioterror attack on the city and its potential spread to other cities involve options of quarantine and vaccination response strategies. Vaccination strategies include mass vaccination (MV), trace vaccination (TV), and city-block vaccination (CBV). A mass vaccination strategy refers to vaccinating 100 percent of the population in all the geographic locations. A trace vaccination strategy relies on actual contacts between the infected victims to trace the spread of the infection and model a vaccination strategy to "chase" the contagion. A city-block vaccination strategy refers to immunizing or treating 100 percent of the population of a particular geographic location in the

Table 2. Pearson correlation coefficients

	Variable	N	Mean	S.D.	Min	Max	1	2	3	4	5	6
1	Quarantine	135	2.61	1.20	1.00	5.00	1.00**					
2	Vaccination	135	2.52	0.98	1.00	4.00	0.46**	1.00**				
3	Intensity	135	2.30	0.80	1.00	3.00	0.63**	0.68**	1.00			
4	Timing	135	2.58	1.05	1.00	5.00	0.49**	0.65**	0.53**	1.00		
5	Number infected	135	244.60	87.83	95	323	0.48**	0.69**	0.70**	0.64**	1.00	
6	Number deceased	135	2.80	1.48	1.00	5.00	0.29**	0.49**	0.48**	0.47**	0.80**	1.00
7	Public mood	135	4.20	0.40	4.00	5.00	0.18*	0.21*	0.23*	0.17*	0.28**	-.3*

*denotes significance at $p \leq 0.05$; **denotes significance at $p \leq 0.001$.

hope of containing or buffering the infection with a "firewall" effect. City blocks are chosen in relation to the outbreak, and each of the blocks is vaccinated every alternate day. Quarantine strategies included the more passive city-block quarantine (CBQ) approach or a military-style, more extreme quarantine (EQ) approach. The city-block quarantine implies quarantining 100 percent of the population in a particular geographic location. The extreme quarantine implies quarantining 100 percent of the population in all the geographic locations.

The intensity of the strategy implementation is correlated with the strategy choice. With a mass vaccination strategy, a low intensity implementation, in relation to the timing of the response, would vaccinate up to 10 percent of the population every day until 100 percent of the population was vaccinated, whereas a high intensity implementation, in relation to the timing of the response, would vaccinate up to 20 percent of the population every day until 100 percent of the population was vaccinated. With a trace vaccination strategy, a low intensity implementation in relation to the timing of the response would vaccinate up to 10 percent of the population every day based on the trace model until 50 percent of the population was vaccinated, whereas a high intensity implementation in relation to the timing of the response would vaccinate up to 20 percent of the population every day based on the trace model until 50 percent of the population was vaccinated. With a city-block vaccination strategy, a low intensity implementation in relation to the timing of the response would vaccinate up to 10 percent of the population of the blocks every day until 100 percent of the population of the blocks was vaccinated, whereas a high intensity implementation in relation to the timing of the response would vaccinate up to 20 percent of the population of the blocks every day until 100 percent of the population of the blocks was vaccinated.

The timing of the vaccination or quarantine strategy implementation assumes no action is taken until the attack is evidenced (days 1 to 4). The response implementation can therefore only take place immediately after the attack (day 5), after slight delay (days 6 to 7), or after extreme delay (day 8). The respective vaccination or quarantine strategy choice is implemented correspondingly. Theoretically and intuitively earlier implementations of vaccination or quarantine strategies are expected to have superior results for contagion containment (Kaplan et al. 2002), but prior literature hypothesizes a trade-off in the timing of the response (Chaturvedi et al. 2005). Responding too aggressively, too early, may adversely impact public mood and produce collateral casualties, whereas responding too late may allow the contagion to grow beyond the control of the responders.

Results

This study makes use of Purdue University's Measured Response simulation environment on the SEAS platform in conjunction with the MR03 training

Table 3. MR03 outcome results

MR03 Results	Day 1	Day 2	Day 3	Day 4	Day 5	Day 6	Day 7	Day 8	Day 9
Number exposed	368	472	570	674	766	826	824	824	1,236
Number infected with symptoms	10	8	7	7	7	6	5	5	19
Number deceased	0	1	2	2	2	3	4	4	5
Cumulative number infected	184	420	705	1,042	1,425	1,838	2,250	2,662	3,280
Cumulative number deceased	0	1	3	5	7	10	14	18	23

exercise held in July 2003 to generate experimental data based on our decision model. Due to agent specific nature in components, additional response data were also collected through a survey instrument administered during the MR03 exercise. An overview of the results of the MR03 exercise is provided in Table 3 and discussed in the next section.

▶▶ Discussion and Conclusion

The discussion, observations, and results of the Measured Response training exercise series in this chapter indicates that, with early intervention and effective communication, coordinated multiagency responses can be formulated and implemented in order to successfully contain a bioterror attack for minimizing health and economic impacts on the population. Throughout the event, officials struggled with the communication process. Quickly noticing the lack of guidelines between and within levels, the importance of a well-defined information flow became evident. This indicates that in the event of a biological attack, it is essential that the right people have the right information at the right time. To ensure that this occurs, policies are needed to streamline the flow of information across levels. This need was discovered early in the simulation, because the diagnosis of the disease could have been done more efficiently had different personnel been informed of the events that were taking place.

While these observations indicate that government officials clearly struggled with the communications and decision-making processes during MR03, much as they did in the prior year's exercise (MR02), this is likely reflective of day-to-day real-life interactions. This may be due in part to a lack of guidelines between and within levels. The need for a well-defined information flow and response framework may be evident from these observations. As discovered through the MR02 and MR03 exercises, a model and operational policy is needed to streamline the flow of information across levels. With such an approach, the diagnosis of the bioterror attack can be done more efficiently when different personnel are informed of the events.

This chapter explored the diverse strategic, organizational, and operational issues resulting from unconventional threats, such as those from nuclear, biological, and chemical weapons. These threats require a significant improvement in our capabilities to model the human behavior. The representation of human behavior must reflect human capabilities, cognitive processes, limitations, and conditions that influence the behavior (e.g., morale, stress, panic). This research stream focuses on the development of computational models of human behaviors to build artificial agents. These active, behaviorally accurate, artificial agents enable realistic simulations of economies, nations, communities, markets, and organizations with only a few people in the loop. Different agents simulate different behavior patterns, leading to novel ways for analyzing and testing policies and theories. While many challenges remain with studying these issues, we hope that through continued training and refinement of this simulation environment and exercise series, coordinated government response capabilities for bioterrorism events can be improved.

▶▶ Acknowledgements

The authors wish to acknowledge the National Science Foundation and the 21st Century grant, which funded this project. We wish to thank the Purdue Homeland Security Institute (PHSI) for the use of their simulation environment and Chih-Hui Hsieh and Tejas Bhatt and the rest of the PHSI team for their extensive assistance with this study during the 2002, 2003, and 2004 Measured Response training exercises, as well as with coding and compiling the data. We also wish to thank the members of the departments of homeland security, health and human services, and transportation from federal, state, and local governments who assisted with and participated in the Measured Response training exercises. We also wish to thank Eric Dietz, Jari Niemi, and Gina Niemi of PHSI for their considerable assistance with this project. We further thank Steve Green and the participants of his 2003 research methods seminar for input on early initial drafts of this project and for his helpful comments on our survey instrument. Finally, we wish to thank Tom Brush, Tim Folta, Roberto Mejias, Kent Miller, Mark Shanley, and Derek Ruth, who have provided informal input throughout the course of this project.

References

Chaturvedi, A., and S. Mehta. 2002. The SEAS Simulation Environment. Technical Reports, Purdue University, West Lafayette, IN., 519–530.

Chaturvedi, A., S. Mehta, and P. Drnevich. 2004. Computational and Live Experimentation in Bio-terrorism Response, *Dynamic Data Driven Applications Systems*, F. Darema, ed., Kluwer Publications: Boston, MA.

Cheney, R. 2002. Remarks on *Fox News Sunday*, May 20, 2002.

Chyba, C. 2002. Toward Biological Security. *Foreign Affairs* 81(3):122–136.

Croddy, E., C. Perez-Armendariz, and J. Hart. 2002. *Chemical and Biological Warfare: A Comprehensive Survey for the Concerned Citizen*. New York: Springer-Verlag.

Gode, D. K., and S. Sunder. 1993. Allocative Efficiency of Markets with Zero Intelligence (ZI) Traders: Market as a Partial Substitute for Individual Rationality. *Journal of Political Economy* 101:119–37.

Gode, D. K., and S. Sunder. 1997. What Makes Markets Allocationally Efficient?" *Quarterly Journal of Economics* 112:603–630.

Inglesby, T., R. Grossman, and T. O'Toole. 2001. A Plague on Your City: Observations from TOPOFF. *Clinical Infectious Diseases* 32:436–445.

Kaplan, E. H., D. L. Craft, and L. M. Wein. 2002. Emergency Response to a Smallpox Attack: The Case for Mass Vaccination, *Proceedings of the National Academy of Sciences*, 6(16) 10935–10940.

Kelly, J. 2005. Death and Destruction, Just for Practice. *Washington Post*, April 13, 2005, C13.

Milgram, S., and H. Toch. 1969. A Note on Drawing Power of Crowds of Different Size. *Journal of Personality and Social Psychology* 13:79–82.

O'Toole, M., and T. Inglesby. 2002. Shining Light on "Dark Winter." *Clinical Infectious Diseases* 34:972–983.

Rvachev, L. A., and I. M. Longini. 1985. A Mathematical Model for the Global Spread of Influenza. *Math Bioscience* 75:3–22.

U.S. Army. 2001. 21st Century Bioterrorism and Germ Weapons. *U.S. Army Field Manual for the Treatment of Biological Warfare Agent Casualties*. Washington, DC: U.S. Government Printing Office.

Yamori, K. 1998. Going with the Flow: Micro-Macro Dynamics in the Macrobehavioral Patterns of Pedestrian Crowds. *Psychological Review* 105(3):530–557.

Index